People the Planet Needs Now

People the Planet Needs Now
Copyright © 2025 by Dudley Edmondson

Published by Adventure Publications
An imprint of AdventureKEEN
310 Garfield Street South
Cambridge, Minnesota 55008
(800) 678-7006
www.adventurepublications.net

Cataloging-in-Publication data is available from the Library of Congress
ISBN 978-1-64755-277-0 (hardcover); 978-1-64755-278-7 (ebook)

Photo credits
Front cover: All photos by Dudley Edmondson

Back cover: Dudley Edmondson photo by Chad Brown;
bottom photo by Dudley Edmondson

Page 2 images under license from Shutterstock.com:
Upper left sewage pipe: **Vastram**; upper center smokestacks: **TR STOK**;
upper right car: **Tricky_Shark**; middle left city barrels: **1000 Words**;
middle left center plastic bottles: **Mr.anaked**; middle right center rusted
barrels: **SvedOliver**; middle right garbage bags: **saravuth**; lower left oil
refinery: **Ssisabal**; lower right pipeline: **Maksim Safaniuk**

All other photos are credited in the captions.

Editors: Brett Ortler, Andrew Mollenkof, Emily Beaumont, and Holly Cross
Designer: Hilary Harkness
Production: Monica Ahlman

10 9 8 7 6 5 4 3 2 1

People the Planet Needs Now

VOICES FOR JUSTICE, SCIENCE, AND A FUTURE OF PROMISE

Dudley Edmondson

PUBLICATIONS

Adventure

an imprint of AdventureKEEN

A MOUNTAIN RANGE INSIDE THE ARCTIC CIRCLE ON THE ATIGUN RIVER, ALASKA
BY DUDLEY EDMONDSON

This book is dedicated to all the tireless champions for environmental and social justice around the world. From the process of putting this book together and other conversations I've had, I know this work is highly stressful, emotionally draining, and, unfortunately, never-ending. This book is a heartfelt thank-you for all the work you do.

RICKY DEFOE, AN ELDER AND A PIPE CARRIER OF THE FOND DU LAC BAND OF LAKE SUPERIOR CHIPPEWA
BY DUDLEY EDMONDSON

CONTENTS

Preface	xii
Foreword	xiv
Ibrahim Abdul-Matin	6
Fatima Ashraf	14
Majora Carter	24
Corina Newsome	32
Roxxanne O'Brien	40
Deja Perkins	50
Queta González	60
Dr. Drew Lanham	68
Chad Brown	76
Rue Mapp	86
Christopher Kilgour	92
Nicole Jackson	104
Alex Troutman	110
Siqiñiq Maupin	120
Nikola Alexandre	130
Tamara Layden	140
Ricky DeFoe	148
Jason Hall	158
Ashanee Kottage	172
Dr. Lorena Rios Mendoza	182
Dr. Sebastian Echeverri	190
Charlie "Mack" Powell	198
Reverend Edward Pinkney	206
Sharon Lavigne	216
A.G. Saño	226
Resources	241
Acknowledgments	245
About the Author	246

**IBRAHIM ABDUL-MATIN
FATIMA ASHRAF**
BROOKLYN, NY

Are we creating
spaces for people
who come after us
to live and thrive?

page 6

MAJORA CARTER
SOUTH BRONX, NY

Our ability to reclaim
our communities is
about us creating a
future for ourselves so
that we don't feel like
we've got to move out of
our neighborhood and
live someplace better.

page 24

CORINA NEWSOME
ATLANTA, GA

If we are not positioning
the people being harmed
as the leaders in place-
based environmental
problem-solving, then
they will continue to
be exploited even in
solutions to the problem.

page 32

ROXXANNE O'BRIEN
MINNEAPOLIS, MN

If you want to know
how much corruption
is in your community,
pay attention to how
much pollution is in
your community.

DEJA PERKINS
RALEIGH, NC

Diversity matters,
including diversity
of experiences and
of thought.

QUETA GONZÁLEZ
PORTLAND, OR

By identifying our
implicit associations,
and how dominant
culture comes through
in our actions, behav-
iors, and values, we can

DR. DREW LANHAM
CLEMSON, SC

Conservation to me is just another word for caring, but caring in a way that saves something for somebody else.

page 68

CHAD BROWN
PORTLAND, OR

I want youth and veterans who are troubled and are displaced, fighting depression, to know that the river is not just there to fish or swim, but it's also a place of healing.

page 76

RUE MAPP
OAKLAND, CA

How do we get people to care about something that they don't have a relationship with? That's the role for Outdoor Afro: helping people develop a trusting, loving, and inspired relationship with the outdoors.

page 86

CHRISTOPHER KILGOUR
MADISON, WI

One of our founding mantras is: Take up space unapologetically.

page 92

NICOLE JACKSON
COLUMBUS, OH

Getting Black women in nature provides a great opportunity to find solutions to mental health concerns, and it creates more joyous and educational moments in outdoor spaces.

page 104

ALEX TROUTMAN
AUSTELL, GA

I truly believe that nature is for everyone, and we should all be able to have the same privileges that our white counterparts have without worrying about having the police called on us.

page 110

SIQIÑIQ MAUPIN
FAIRBANKS, AK

The Arctic is warming at a fast pace, and corporations have so much to gain from telling us everything is fine.

page 120

NIKOLA ALEXANDRE
CAZADERO, CA

It's a failure of imagination to not realize that human beings can be more than destructive, but that they can be generative to the land.

page 130

TAMARA LAYDEN
FORT COLLINS, CO

I want to help Indigenous communities use ecological and Western science tools—essentially use colonizer tools against the colonizer—to increase Indigenous sovereignty and propagate Indigenous stewardship.

page 140

RICKY DEFOE
CLOQUET, MN

The wise ones say that without justice, how can you have reconciliation? Justice means action.

JASON HALL
PHILADELPHIA, PA

Our conservation efforts need to be considering how the Indigenous folks of this land were managing it before Europeans arrived.

ASHANEE KOTTAGE
WASHINGTON, D.C.

If you think about how many Indigenous people and minoritized people were displaced to create national parks, they have a tragic history.

DR. LORENA RIOS MENDOZA
SUPERIOR, WI

Wherever you have humans, you have plastics.

page 182

DR. SEBASTIAN ECHEVERRI
SOUTH ORANGE, NJ

The real world is complicated, and real biology is complicated and is best understood from a diverse set of perspectives.

page 190

CHARLIE "MACK" POWELL
BIRMINGHAM, AL

I am the founder and president of P.A.N.I.C., which is People Against Neighborhood Industrial Contamination. What we want is justice.

page 198

REVEREND EDWARD PINKNEY
BENTON HARBOR, MI

The only thing we want is safe, clean water for this community.

SHARON LAVIGNE
ST. JAMES, LA

We need attorneys to help us. We need legal help. We need legal representation for the people. I want it for the people whose loved ones have died because

A.G. SANŌ
MANILA, PHILIPPINES

Basic human rights are trampled because of the changing climate and the activities of corporations and governments around the world. No one is

DUDLEY EDMONDSON
BY STEVE ASH

PREFACE

It is an honor and privilege to share this collection of personal reflections from Black and Brown storytellers with global origins and perspectives. Within the pages of this book, they discuss their lives. Some are scientists, some are activists for environmental or social justice, and many are both.

These 24 viewpoints were gathered through several hours of conversations over Zoom with each individual. The stories before you are just a small portion of those conversations and represent what I felt were the most thought-provoking viewpoints. I intentionally left each chapter in first-person perspective to preserve each person's unique voice. Sentence structure in the way we speak and write are not always the same, so this book is intended to feel more conversational and less formal in terms of style. I want readers to feel as though they're having a one-on-one dialogue with each storyteller. If I were to interview people and reinterpret those conversations and add my perspective, it would feel disrespectful and extractive. It is very important to me to preserve the emotions and to let each story remain unfiltered.

So many people were generous enough to share deeply personal things about their lives. I asked everyone about their childhood, as I believe it makes a familiar point of connection for readers. We all have that in common, and sometimes reflecting on those formative times in our lives can help us understand how we became the people we are today. A few people were even courageous enough to share some of their physical and mental health struggles. Their hope was that by doing so, they might help others dealing with similar issues.

Nearly all the storytellers spoke about race or racism. This topic influences everything, from science and what is considered knowledge, to where you might live in a particular city, and even the quality of healthcare and education you might receive. Needless to say, people of the Global Majority understand that racism is not just an American problem, but it exists around the globe, as global environmental policy determines the quality of life for people around the world.

This book offers a rare opportunity to see and hear directly from BIPOC scientists and activists about the problems they have witnessed with our current methods of addressing everything from climate change to industrial waste and even how we design our cities. Unfortunately, Black and Brown people are often most directly affected by these problems, so their unique insights can offer the possibility of new solutions.

Black and Brown people around the globe have always had an interdependent relationship with nature. In fact, in many cases, they were crucial to the health of the environment because they created cultural traditions that acknowledged stewardship of the land because of their longstanding belief that humans and nature are one.

Perspectives from people of the Global Majority can help us rethink our relationship with nature and push for international agreements that view the natural environment and people equally, so that all living things can continue to call this planet home.

—DUDLEY EDMONDSON

SHELTON JOHNSON
BY ROXANN MULVEY

FOREWORD

Years ago, I rode on horseback in the Martin Luther King Jr. Parade for the city of Merced, California. I wore the uniform of a U.S. Army Cavalry Sergeant circa 1903, the same attire that was worn by African American troops who served as de facto *National Park Rangers* in Yosemite and Sequoia National Parks around the turn of the last century. My period McClellan saddle, saddle bags, bridle, reins, blanket, etc., were also tangible reminders of a military legacy that had nearly been forgotten for over a century.

I wasn't the first ranger to share this untold story with the public. Ranger Althea Roberson, the first African American woman to work as a park ranger in Yosemite, and Ranger Kenneth Noel preceded my contribution by decades. There's a saying that it takes a village to raise a child. Well, it also takes a community of people to raise the profile of any legacy that's not about a Euro-American male of property and means. It's important to note that, although women were *not mentioned in the original US Constitution,* that exclusion did not mean they were not "free persons." It also didn't mean that they held the same status as Euro-American men. After all, that's what was meant by the phrase *We the People.* Race, gender, class, sexuality, and other factors always influence the dynamic of why some stories are remembered, even celebrated, while others are forgotten.

Early on, I learned that the best remedy for an untold story is amplification—and that the greatest amplifier is the media: radio, newspapers, magazines, television, and film. I partnered with Yosemite's Chief of Media Relations, and he in turn began to help spread the word about my work to tell this untold story, and the rest is literally history.

This is why Dudley's book is so important, timely, and necessary! After I rode in that MLK Parade in Merced, I met a teacher from one of the town's public schools. She told me that she spoke to her class about my work to welcome underrepresented groups into *America's Best Idea,* and a young African American male student responded by saying, "*Black people

have nothing to do with national parks!*" or words to that effect.

This young man, whom I never met, was absolutely certain that national parks and African Americans resided in different universes, yet he lived in a gateway town to one of the most iconic national parks in the world, a sacred natural space that has inspired international tourism for over 100 years, and a UNESCO World Heritage Site that was protected by hundreds of African American cavalry and infantry over a decade prior to the creation of the National Park Service in 1916. It was, and continues to be, the *homeland of multiple indigenous cultures* whose roots here in the Sierra Nevada extend over 10,000 years back into antiquity.

This is why *People the Planet Needs Now* by Dudley Edmondson is both important and timely and also necessary for there to be a future for us all that's worth inhabiting. There are just too many young people from incredibly diverse backgrounds who have no idea that not only are our national parks protected public and *sacred* lands and part of a shared inheritance, but that these spaces also provide a profound stage for both healing and the good health that comes from being physically, psychologically, and spiritually connected to the world beneath our feet, the mother of us all.

The human beings profiled in this book are engaged in work that will lift those up who feel weighed down; hand a lantern to those walking through darkness; provide a compass to neighbors just trying to get *home*; and, ultimately, offer by the example of lives lived well how we all can benefit by working together on behalf of something so much greater than ourselves, national or cultural identity, or politics.

Yes, we can all work together to the lasting benefit of all our relations by freeing ourselves of illusions of ownership of anything other than our own spirits and simply, and not so simply, being fully present in this world, as we move forward together to a place where we are all held and embraced by that wildness that precedes and follows our own awakening—and the fierce grip that never weakens as we all fall into the heart of a dawn that has no ending, illuminating a *world we have yet to truly inhabit and share with all other living things.*

The steps are laid out before us, with each step being a life well lived, and living still within every story that follows. All you must do to begin this journey is to simply open a window or a door, or, most magical of all, turn the pages of this book!

—Shelton Johnson,
author of *Gloryland*

Race, gender, class, sexuality, and other factors always influence the dynamic of why some stories are remembered, even celebrated, while others are forgotten.

—SHELTON JOHNSON

People the Planet Needs Now

ENVIRONMENTAL ISSUES AND POLLUTION DISPROPORTIONATELY AFFECT BIPOC PEOPLE.

THIS BOOK IS NOT SUPPOSED TO MAKE PEOPLE FEEL GREAT

It's supposed to piss you off just a little bit. Hopefully, it will make you ask questions about companies and society. The interviews in this book are full of injustices that I think will make people think, "What the hell? How can corporations, industries, or governments do this?! And to think nothing of the Black and Brown people so often at the center of these environmental issues. How is this happening in America today?" It is hard to understand how people can turn a blind eye to those kinds of things and let people suffer. But these are voices for justice, and if we listen, a future of promise and hope awaits.

—DUDLEY EDMONDSON

A.G. SAÑO PAINTING A DOLPHIN MURAL WITH COMMUNITY VOLUNTEERS
COURTESY OF A.G. SAÑO

The story belongs to the storyteller.

—DUDLEY EDMONDSON

IBRAHIM ABDUL-MATIN
URBAN STRATEGIST, AUTHOR, AND ENVIRONMENTAL ACTIVIST

FATIMA ASHRAF
POLICY ADVISOR AND PROFESSOR OF PUBLIC POLICY

BROOKLYN, NY

Ibrahim and Fatima are a power couple in New York known for their significant contributions to social and environmental justice. They embody a deep love for community, family, and environmental stewardship. Through unique experiences and global perspectives, they advocate for forgotten communities, challenge cultural vilification of Muslims, and inspire action within the Muslim community toward conservation and faith-driven responsibility toward the land.

—DUDLEY EDMONDSON

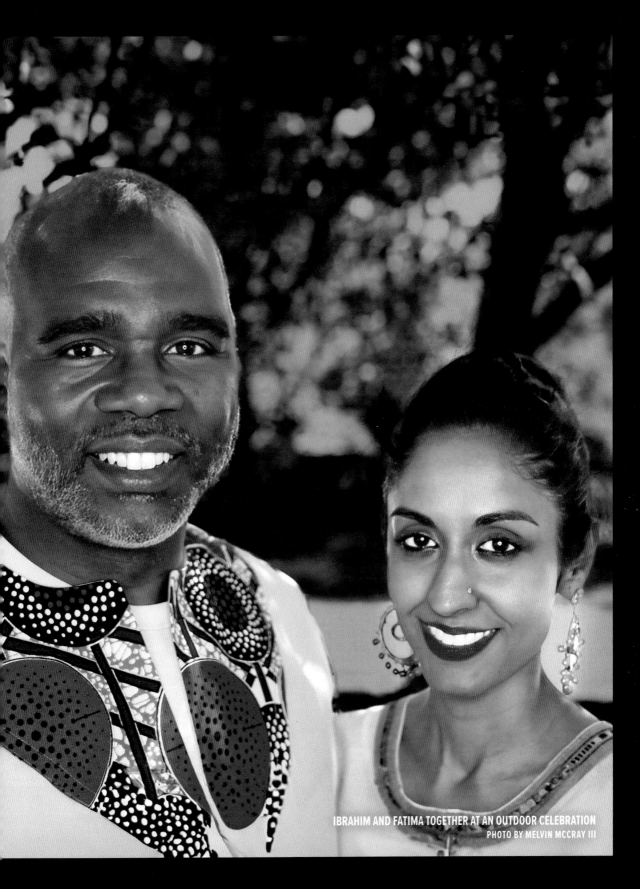

IBRAHIM AND FATIMA TOGETHER AT AN OUTDOOR CELEBRATION
PHOTO BY MELVIN MCCRAY III

BUILDING THE FUTURE

I always give myself a title. In the past, I've called myself a translator. These days, I call myself an urban strategist. I am constantly thinking about how we make decisions around water, waste, energy, food, and transportation. How did we solve these problems in the past? How do we solve them in the present? And how will we solve them in the future?

I help people, organizations, governments, researchers, and academics think about the built environment, energy, water systems, storm management, and the resiliency of our infrastructure. I help startups with new products and technologies and how they can be brought to scale. I push everyone to think critically. My favorite question to ask these days is if someone from ancient times showed up today, and they were looking for water and a place to use the bathroom, what would they say about our systems, their ease, their usability, and longevity?

MY CHILDHOOD GROWING UP IN BROOKLYN

My family is beautiful. My father and I have a wonderful relationship. We talk about many aspects of our lives, and I run every major decision I make by him. My mom too. They have always been my emotional, spiritual, and cultural anchors.

My parents exposed me to beautiful things, despite raising me during the crack and crime epidemic of the 1980s in New York. It was a war zone. There's no place in the US today with that volume of guns and drugs. To be a kid in that moment imagining my future life, I would have never dreamed of what I have now, where I've been, who I've met, and what I've done. I still

A NEWSPAPER CLIPPING OF IBRAHIM AS A FOOTBALL PLAYER WHEN HE LIVED IN UPSTATE NEW YORK
COURTESY OF FATIMA ASHRAF

feel like I've left a lot on the table. I still feel like I have a lot to do.

I come from Black Muslim, militant, segregated Brooklyn. We thought the revolution was coming back then. I remember the first time that I saw a white person. I asked my dad, "What's wrong with him?" He said, "I'll tell you later." We were so protected from the often-harsh realities of integrated places. That was a blessing. My understanding of the universe was incubated in this way.

After my parents' divorce, my siblings and I moved to upstate New York with our father. It was all white, but not rich white. It was country white, farmer white. I found out later in life that many of the dads were involved in the crack and cocaine trade. These were poor, rural, white folks who worked hard, played lots of football and basketball, and ran track. During the summers, we were back in New York City at the African Street Festival, surrounded by all the beautiful cultural expressions of Blackness and African Diaspora. It doesn't exist in this same way now because of police presence and gentrification. The upstate/downstate dichotomy colored (literally) every aspect of my life. We were creating our own thing back then.

I JUST DON'T WANT TO LEARN HOW TO KILL PEOPLE

I received a football scholarship to attend the University of Rhode Island. But I had a number of different offers from Cornell, West Point, Syracuse, and Harvard, among others. I rarely tell the West Point story. I was in my junior year; the recruiter was in my house, papers on the table. I had the pen in my hand. My dad, the recruiter, and I just stared at the pen. After a few minutes that felt like hours, I put the pen down. "I can't do this." The recruiter asked, "Why?" I said, "I just don't want to learn how to kill people." I remember thinking, maybe I could kill someone. But not for this country, not for this flag. Maybe if someone broke into my house and I had to defend myself and my family. I couldn't stomach being sent around the world by the US military, listening to whoever they branded as the bad guys.

I took the scholarship to the University of Rhode Island. When I visited the campus, it was next to the ocean. How could I be anywhere in the landlocked world after seeing the sea? I bring up the fact that I received a full scholarship because my dad told my older brother and me when we were in middle school, "I can't pay for you two guys to go to school. The money I have is for your three younger sisters. You guys have to figure it out on your own." My brother and I made a deal late one night. We decided to get full athletic scholarships. He got track, I got football. This meant, and this is important, that we had zero debt coming out of college.

In Islam, debt is haram or impermissible. You're not supposed to be in debt. I didn't have the weight of debt on my back after graduating, which was a good thing because a few years later we landed in the worst recession since the Great Depression. Without debt, I was free to follow my heart, my interest, and my values. I decided back then that I didn't want to be rich because it would force me to compromise my values. I didn't want to make money by compromising my principles.

I REALLY WANTED TO UNDERSTAND HOW CITIES ARE RUN

Almost every decision that I've ever made in my life—where to live, where to work, even who to marry and how to raise my kids— has been connected to my commitment to protecting the planet, my love of community organizing and activism, and my desire to work with young people.

My commitment to protecting the planet has been the most

consistent theme in my life. One of my first jobs right out of college was at Prospect Park as the Director of Youth Programs. My job was to help young people build a connection with land, stewardship, and planet Earth. That's when my mindset started to shift toward civic engagement, and how we make decisions about water, energy, transportation, and food infrastructure. When I decided that I needed to understand how cities are run, I decided to get a master in public administration. I was less interested in policy and more interested in execution. This frame helped me articulate the outline for my book, *Green Deen: What Islam Teaches About Protecting the Planet*. What I learned in graduate school also kicked other things into motion, including meeting my wife, locking that down, children soon after, and everything since.

In 2010, I started work in the Mayor's Office of Long-Term Planning and Sustainability. I wasn't a buttoned-up guy. But in this job, you couldn't go to a meeting without a jacket and tie. It was a new experience for me. In that job, I was always thinking 100 years into the future. What is the city going to need? What will be its challenges? How will we address

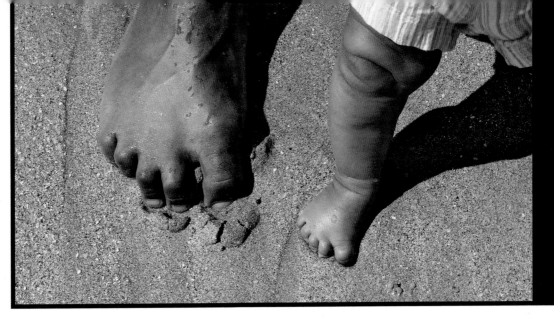

IBRAHIM PLACING HIS FOOT NEXT TO HIS SON'S IN THE SAND ON A BEACH
COURTESY OF FATIMA ASHRAF

these challenges? A few years later I took an incredible position in the Department of Environmental Protection, the agency responsible for NYC's water. I was able to focus on these same questions about the future, but specific to water, water systems, water availability, storm water, and more. As a consultant, I've advised businesses, nonprofits, public and private entities on how to create, evaluate, and implement low-carbon, low-emission policies and strategies—all designed to build our collective future in the least harmful way possible.

ALWAYS SETTING A GOOD EXAMPLE FOR MY THREE SONS

My life is entirely oriented around my sons. I am about to cry just thinking about them. Everything I do is for them, to make sure that they have a good example. Children don't do what they hear or what they are told. They do what they see. I'm one of the dads in the stands at all of their sporting events. I cart them from school to soccer. I carry them on my back and shoulders, sometimes all three at the same time. Everyone knows me, and everyone knows my sons and that I am their father.

Really quickly—about baseball. It's a very family-oriented sport. I think it's important for a couple of reasons. First, it's an all-American sport. I am definitely not all-in on the American experiment. In fact, as a Black man and an environmentalist, I simply cannot be. But I want my kids to have the cultural currency that comes with being baseball players. And, more importantly, baseball is the sport that teaches them the most about failure. You only do better with practice. You put in your reps, you put in time, and you see it. It makes you better. But no matter how great you are, you'll strike out or get out. It's baseball.

> Everything I do is for them, to make sure that they have a good example. Children don't do what they hear or what they are told. They do what they see.

Setting an example for my sons through my practice of prayer is also essential. In Islam, we pray five times a day. There are also prayers that happen in the middle of the night. In Islam, we believe God is closest to you at night. I remember waking up as a child and seeing my dad pray. He didn't know that I saw him. That always stuck with me. My whole focus is on setting an example, so if I were to die tomorrow, my boys would have a template of how we are as men in our family. They would have a good chain of transmission of information from post-slavery to now. Those things are important to me.

IMPORTANT TRUTHS

I'm always wrestling with existential questions. Lately, I've been wrestling with the truths of the Black American experience and the truths of the slave trade. I've always learned about slavery as something that was "done to us," that we were snatched out of Africa by the European colonizers. But before the colonizers came, we were living this idyllic life.

Recently, I've started to understand that may not be the whole story. Before the colonizers arrived, we may not have been as unified as we were taught. The truth is that Africans also sold their African brethren into slavery. The colonizers divided and conquered us on the continent, and that contributed to how swiftly the slave trade occurred.

These truths are important to me because I'm not trying to raise my sons with false narratives. The empowering part of the true narrative, to me, is that if we are united as Black people, we are literally the most powerful people on the planet.

So much was taken from Africa. People were taken first, then, in the late 1800s, with the Berlin Conference, they sucked Africa dry of resources and controlled what was left. Can you imagine if we were united and prevented all of the extraction and exploitation? The cancer of capitalism may not even exist right now.

BLACK UNITY IS THE IMPERATIVE

I think unity is the imperative for Black folks. This is why I get so frustrated with popular culture. There's so little unity. Even this "versus" thing where they pit R&B artists against other R&B artists. It's part of the re-creations of these false narratives—Malcolm vs. Martin, Booker T. Washington vs. W.E.B. Dubois. All of these narratives highlight how we were divided and conquered from the beginning. With my sons, I want them to know these existed and then move them to a narrative of unity. We are Black folks that lean in. When you're in an all-white room, and you see a Black man or Black woman, you lean in. Be like, "What's happening, brother? What's poppin'? What's your story?"

The challenges that we face right now are dramatic, and we are not in a good state. Everyone wants to blame another. I'd like to make sure us Black folks are on the same page so we can overcome these challenges.

Islam and other religions teach that life is a test for everyone. I don't want the white man's test because of what they have done under white supremacy—those are dramatic tests. If they're going back to God with colonization and imperialism and capitalism and so much blood on their hands—that's tough. We have different tests. I think our test is—do we support each other? Are we creating spaces for people to live and thrive who come after us? Are we managing our resources, our systems, so that people later have it better? It's what our ancestors tried to do in the past.

LEARN MORE

Green Deen: What Islam Teaches About Protecting the Planet,
Kube Publishing, 2012

AUTHOR'S NOTE: Ibrahim died suddenly, unexpectedly, and tragically on June 21, 2023. He died doing what he loved most—taking care of his three boys. Ibrahim inspired everyone who came into contact with him. His unique ways of looking at how we solve persistent human problems was motivating for people of all ages, from all walks of life. His book, *Green Deen: What Islam Teaches About Protecting the Planet*, is a guide on how to be better world citizens and better environmental stewards, regardless of your beliefs.

FATIMA ASHRAF

(Ibrahim's Widow)

**POLICY ADVISOR
AND PROFESSOR
OF PUBLIC POLICY**

COURTESY OF FATIMA AS

IBRAHIM WAS PREPARED FOR DEATH

His heart stopped. It was very sudden, and it was very immediate. The explanation that Ibrahim would give you, if he was able to speak from the other realm, is "Allah decided that it was my time." Personally, this explanation is hard for me because it leaves me with so many whys. Why when the boys are so young? Why when the world is so tough? Why?

I know in your interview with him, he reflected on his childhood in crack-era Brooklyn and in rural upstate New York. As he said, he could have never imagined as a 7-year-old that he would have gone to so many countries, written a

book, been on speaking tours, and worked at high levels of government. He would have never imagined that as a child.

But he did imagine dying, all the time. He wrote about death in every journal, in nearly every essay. I have not come across a speech yet that he has given, where he has not mentioned the idea of death. I don't think a lot of people prepare. Ibrahim was really preparing. He knew that this life is a fleeting moment and that what's to come is infinite and hopefully better, so he would say, "I'm going to do what I got to do. I'm going to do the best that I can do here to make there [the afterlife] the best that it can be."

IBRAHIM'S DEATH

On June 21st, Ibrahim was making lunch for the boys. The younger two were with him and they said they wanted to go outside to play soccer. They came back, maybe 10 minutes later, and found him on the floor of the kitchen. The boys have an iPad with everybody's phone numbers, so the first person they called was my younger sister who lives in Seattle. They FaceTimed her and said, "Dad's on the floor. He's not waking up." She immediately called 911 from Seattle and told the kids to wait in the driveway for the ambulance. Then she jumped on a plane.

I was also on a plane. My siblings were calling me. I couldn't swipe to take a call while in flight, so my brother texted me, "Something's wrong with Ibrahim. There are paramedics at your house." Of course my heart sank, and I started shaking. When the plane landed, I called my brother, who was on the phone with the paramedics and the ER doctors. They confirmed my beloved Ibrahim had collapsed. They tried chest compressions. But he was not to be saved. I didn't leave the airport. I got on a flight back right away.

When I saw Ibrahim in the hospital, he looked 15 years younger and had a light smile on his face. I am so happy that I was able to commit his gorgeous face to my memory. I sang to him, prayed over him. Then they released him to the mosque who would arrange for his ghusul, the Islamic washing, and his janaza, the Islamic burial.

I declined an autopsy because he never wanted one. He always said that he wanted to be buried in perfect adherence to Islamic principles. So that's what I did, and that's why he is buried in California and not at home in New York. We were in our last few days of an 18-month stay in Los Angeles. In Islam, the best possible case is that you are buried within 24–48 hours in whatever place you died. We believe that once we die, our bodies go back into the ground, into the earth, and our souls live on. The soul is not restricted to any place on Earth. So, I felt comfortable burying him in California and not in New York. Somebody said the most beautiful thing to me during the funeral. I wish I could remember who it was. They said, "Ibrahim, like the sun, rose in the east and set in the west."

IBRAHIM WAS A VERY HOPEFUL PERSON

Ibrahim had a childlike innocence. Everybody will agree to this. He hated when I used to jokingly tell people that I have four boys. But it's true, and it's because of his child-like innocence. He had wonderment about the world and all living things and even nonliving things. I'm building his life archive now for our boys, and looking at his Instagram photos is so funny because he would take pictures of the Metropolitan Transit Authority guys digging up potholes on the streets of Manhattan. He was in wonderment about the pipes underneath the sidewalk and, in the same breath, in wonderment about the mountaintops and all of nature. Childlike innocence and wonderment were his defining qualities. That's what made him so easily predisposed to spirituality and faith, belief in the unseen, and also hope, hope for people being better, the world being better. He was a very hopeful person, but also realistic.

I think that Islam gave him both hope and realism. As Ibrahim always said, "Islam is a framework for living, solving problems, using resources, managing relationships, and even figuring out where to live." To him, Islam provided this realistic set of guidelines that he could lean on to guide his own life. It also provided him hope for what's to come. If you do what you're supposed to do here, then the hope for what's to come is a wonderful forever.

I re-read *Green Deen* in preparation for this conversation. One thing I reflected on is how much people struggle with the purpose of life. So many people struggle with the questions of "Why are we here? What are we supposed to be doing?" I think that Ibrahim had these questions figured out. He had a very, very visceral understanding of life's purpose, and it's all encapsulated in the last line of his book, which is, "We praise our creator, take care of our planet, and take care of each other." Now, he struggled with his role in making that purpose happen, so he's had many titles through the ages, and I think that's what he struggled with—"I know the purpose of life. What is my role in fulfilling that purpose?"

THE HANAFI HUNDRED

We met in 2007 at an art opening. He always said that he knew immediately that I was the one. It took more time for me. We started to see each other at Muslim community events, political gatherings like protests and debate-watching parties during the 2008 elections. It wasn't until 2009, when I saw Ibrahim perform onstage, that I fell for him. It was called, "Community Café" and was a night of music and poetry by incredible artists. It was held at the Malcolm X and Dr.

IBRAHIM, FATIMA, AND THEIR SONS VACATIONING IN EUROPE
COURTESY OF FATIMA ASHRAF

Betty Shabazz Education and Cultural Center in Harlem, the site of the old Audubon Ballroom where Malcolm X was martyred.

I was standing in the back with a friend, watching this gorgeous man onstage. I asked her, "Who is that guy?" She said, "That's Ibrahim, the environmental guy, the one who rides his bike around town." In disbelief, I said, "That's not him. I know that guy. He's been trying to talk to me. That's not him." At the end of the night, all these women rushed the stage trying to talk to him, and I thought, "Shoot, I can't believe that's the same guy." So, I elbowed my way through the crowd and made it to the stage to

catch his eye and say, "Hey, it's me! Remember?" And he said, "Yeah, I remember you, come here and meet somebody." He took me to meet his mom who was sitting in the front. I laugh now—I was so vain as a youth!

I praised him at the event and asked him to include me in the planning of the next one. He said that he was traveling for a few weeks but that we should have coffee when he got back. Like clockwork, he reached out when he got back, and we went to a coffee shop on Flatbush and 7th. It's not there anymore.

I come from an immigrant Muslim family, and I was raised to be a practicing Muslim girl. Dating for the sake of dating is unacceptable. The goal is always marriage. So, on our first "date," I brought the "Hanafi Hundred," a list of 100 questions that Imam Abu Hanifa, an ancient scholar of Islam, created for people to consider before marriage. It's basically a compatibility test. I arrived at the coffee shop first. I had printed the Hanafi Hundred and brought pens. When he arrived, I asked him before he even sat down, "Hey, are you in this for the long haul? Is your goal to get married?" Leaning with his hands firmly on the table, without even a microsecond of hesitation, he said "YES." So, we sat for four hours and went through every single question. I told my parents about him soon after that coffee date, and they weren't messing around. It was a quick courtship and then marriage and then kids. We dove into our lives very quickly.

IF I AM MUSLIM, THEN I NEED TO BE AN ENVIRONMENTALIST

Green Deen was very much a part of our courtship, even though he'd developed the idea ages before. Ibrahim's connection to environmental justice came through his practice of Islam. He believed deeply that our purpose in life as a creation is to praise our creator, take care of our planet, and take care of each other. That's our purpose. Islam is very green as a faith. It's all about nature in God and God in nature.

The book's foundation is based on six principles that were developed by Faraz Khan, one of the OG Muslim environmentalists. The first principle is Taqwa or the oneness of God. The second principle is Ayat or the signs of God. We have our Quran, our Holy book, which has six thousand and two hundred Ayats or lines of text. God repeatedly says, "Here are my signs." The signs are the sun, the moon, the mountains, the oceans, the foliage, the animals, the universe, the star paths, the celestial orbits, the circle of life, the reproductive cycle, the nature of people. The Quran is magnificent because it lays out so much about the world, about God's creation, that we've taken centuries to come to know through scientific study.

Ibrahim always said, "When you take a walk in nature, you're reading the Koran, and when you're reading the Koran, you're taking a walk in nature. They are one in the same." Ibrahim learned this concept from his beloved friend Mohamed Chakaki, who is an incredible visionary, scholar, and all-out beautiful human being too. The third principle is Khalifa or

IBRAHIM AND FATIMA POSE WITH THEIR SONS AFTER A SOCCER GAME.
COURTESY OF FATIMA ASHRAF

steward. The whole point is for us to be stewards of this place, of this creation that we have been given. The fourth principle is Amana, the trust that God has given us as creatures with the capacity to reason and make decisions, decisions that allow all of creation to flourish. Then the fifth and sixth principles are Adl and Mizan. Adl means justice and Mizan means balance. So, we're basically the stewards entrusted with creation to do justice and to establish balance. That is the core of our faith, of Islam.

For Ibrahim, being Muslim meant that he must be an environmentalist. And as an environmentalist, he found it easy to be Muslim. Taking care of the planet, the animals, making sure resources are used in just and balanced ways, these are all things God requires of us in Islam, and these are all things that every environmentalist must do. In Islam, we do not believe in scarcity. We know that the earth can regenerate. Everything grows, everything regenerates, even the water cycle. God created it all to be regenerative. Ibrahim took this belief to heart. It's why he pushed so hard for balanced management of resources and systems because he knew from Islamic teachings that just and balanced management over extraction, exploitation, and consumption would allow all of creation to live in harmony.

Of course, capitalism requires scarcity in reality and in mindset. It makes things hard when other traditions that were influenced by capitalism lead people to believe in scarcity and that we have to fight over resources. When some hoard and destroy instead of share and replenish. *Green Deen* had already been developed in Ibrahim's mind, and he was thinking about how to engage the Muslim community in the climate change conversation by showing everybody that our Deen, our faith, our way, our path is inherently environmental. It gives us the framework to live in a way that is harmonious with planet, people, and all of creation.

Ibrahim's loss—to people and planet—is immeasurable. But his book will live on. In Islam we have this concept of Sadaqa Jariya, which are good works that you do during your life that continue to benefit people and the planet after you die. Planting trees, building wells, and writing books are examples of Sadaqa Jariya. I ask all readers to please pray that Ibrahim's Sadaqa Jariya continues to bless the people of this world and that he reaps the benefits of his work in the afterlife. I also ask the readers to pray for our sons, that the loss of their larger-than-life father doesn't diminish their light, but that his memory helps them become giants in their own right. And may we all be reunited in a better place. Ameen.

A beautiful couple—highly educated and deeply immersed in their cultural beliefs. Their lives are what many of us wish for: a strong, tight-knit family, doing amazing things in their community and living and traveling the world with their beautiful sons. Ibrahim was a visionary thinker, a person who could motivate people to reimagine the world. His mission in life was to help people think differently, to get them to reimagine their relationship with the natural world, each other, and make the world a better place. As-salamu alaikum, brother Ibrahim. —DUDLEY EDMONDSON

IBRAHIM IN WONDERMENT OF A BRIGHT WINTER DAY
COURTESY OF FATIMA ASHRAF

LOW-INCOME AREAS ARE OFTEN SUBJECTED TO TOXIC AIR.
BY TRICKY_SHARK/SHUTTERSTOCK

WHAT?!

"The health burden of air pollution is higher for those living in areas of poor air quality. Residents of low-income neighborhoods and communities may be more vulnerable to air pollution because of proximity to air pollution sources such as factories, major roadways and ports with diesel truck operations."

—U.S. ENVIRONMENTAL PROTECTION AGENCY

MAJORA CARTER

URBAN REVITALIZATION STRATEGIST

SOUTH BRONX, NY

Majora Carter's is a significant voice on economic and social uplift by way of the natural and built environment. She works to normalize local property and business development in communities to help retain the local talent who often leave. Her book, *Reclaiming Your Community: You Don't Have to Move Out of Your Neighborhood to Live in a Better One*, highlights the value of nurturing one's neighborhood economically, socially, and environmentally.

—DUDLEY EDMONDSON

MAJORA WALKS ACROSS THE STREET IN THE BRONX, NEW YORK.
BY ELVIN

CHANGING THE NARRATIVE

As a strategist, consultant, and real estate developer, I work on community-based projects or with organizations interested in creating a more environmentally and/or economically developed approach. All the projects that I've done, whether green infrastructure, job programs, or commercial development, were designed to keep people from thinking about leaving the community. Things that we started, like the Boogie Down Grind Café, are the kind of things that make people feel good about being in the community. That's what I want to do, and that's the way I look at my work. It makes people feel like, "I should give my community a second look" or "Why am I listening to all these voices telling me that there's something wrong with being here?" I want to change the narrative about what our communities are. The only way to do that is to transform our communities into what people want them to be, which is something healthy and healing.

GROWING UP IN THE SOUTH BRONX

I was a weirdo kid. Very, very socially awkward. Some things haven't changed. I spent a lot of time on my own in my backyard. It was mostly concrete, but there were raised beds around the perimeter of it, where the Italian family that lived there planted rose bushes and a grapevine. There were all sorts of little roly-poly bugs and other insects and flies. I loved playing in the small garden. That was my me-time and I'd make up imaginary stories. As a kid, I remember being in the street playing with other kids. We felt like the streets and the sidewalks were ours after school. We played Double Dutch, skelsies, and hopscotch.

Those were also allegedly some of the worst days of South Bronx history because of years of financial disinvestment in our community. Landlords were torching their buildings to collect insurance money instead of reinvesting

MAJORA AT 4 YEARS OLD, TALKING TO SANTA CLAUS ON THE PHONE
COURTESY OF MAJORA CARTER

MAJORA'S PARENTS—TINNIE JOHNSON CARTER AND MAJOR WADE CARTER—IN THE 1940S
COURTESY OF MAJORA CARTER

because they couldn't get a loan because of redlining. Despite all that, we were still able to maintain childhood and do the things that kids do outside in our streets. There were great moments that I will treasure. But even I could tell there were problems, in part because of what I saw on TV and how it represented my community to the rest of the world. According to the news, we were the poster children for urban blight. Nobody would know that the nice memories I had of my community ever happened because, supposedly, we were all pimps, pushers, and prostitutes and killing each other. Well, that did happen, clearly, but it was not all that happened. But that's not the way it was ever presented. I mean, presidents would come and see us because we were so bad. They would walk around and be like, "I'm

AFTER A BIG CLEANUP, MAJORA VISITS HUNTS POINT RIVERSIDE PARK IN 1988.
COURTESY OF MAJORA CARTER

going to fix this . . ." Reagan visited, and I remember Jimmy Carter too. I was a little more generous toward Carter. He did seem like he tried. But Reagan, I'm thinking, "No you ain't, get out of my face."

GOING HOME TO FIND MY PURPOSE

My experiences influenced the feeling that I needed to do what was often expected of smart kids like me. I used education to escape my community. We Americans love that Cinderella story where people grow up in a rough-and-tumble neighborhood, get some education, and then get out and be somebody. It was great that I got a chance to sit and stew in the idea that there is something really wrong with communities like ours. It moved me to take action and change that narrative.

After coming back home for graduate school and staying with my folks, I learned the city and state were planning to open yet another huge waste facility on our waterfront. That's when I realized, this is happening because we're poor and of color and politically vulnerable. That is what happens in communities like ours because, statistically speaking, economically disempowered people are easier to push around. And I know that's a very controversial statement, and people don't like it. But let's be real, poor people are much more politically vulnerable. And then you add the fact that being poor has been criminalized.

I don't know what hit me. I think it was God opening my eyes to see things that I hadn't seen before. I remember thinking, I could leave again, and I would just be part of a long line of people from communities like this who leave, never to return. But I wanted to stay and do something about it, in large part because it really pissed me off that we were being treated like that and that our communities were somehow less-than, and by association, we were too.

ENVIRONMENTAL JUSTICE BEYOND ADVOCACY

Starting Sustainable South Bronx was the natural progression for me because I wanted to focus on project-based development and not just straight advocacy. I felt that people, especially in our communities, needed something to look forward to. It didn't come down to what we are fighting against, which I find a lot of advocates focus on. I thought, what are we fighting for? What kind of community do we want? What's going to make us happy right where we are? We worked to create physical projects that change the idea of what our communities look like and how people thought about their environment. Things like transforming dumps into parks. Or advocating for a highway to be decommissioned so that it could be transformed for community use. Or green-collar job training and placement systems, which gave people personal and financial stakes in improving their communities. Frankly, it's the placement part that made us so successful because we worked with future employers to make sure that we were training our folks appropriately so that they could stay in those type of jobs.

RECLAIMING OUR COMMUNITIES

Our push toward supporting the development of more economically diverse communities started with people from inside the community. We've never had a shortage

> I learned the city and state were planning to open yet another huge waste facility on our waterfront. That's when I realized, this is happening because we're poor and of color and politically vulnerable.

of amazing people being born and raised in low-status communities. What we have is a shortage of them staying and reinvesting in these communities. I talk about it in my book *Reclaiming Your Community*.

When you start to see things like doggie daycares and cute cafés where they were not before, that's not when gentrification starts. Gentrification starts when people in that community, long before you see that stuff, start to believe there's no value in those neighborhoods and act accordingly.

How many people know folks come from communities like ours who owned a piece of property and the second their parents died they sold that house to a predatory speculator? We watched from afar as that neighborhood changed and suddenly that house is worth a lot of cash.

Our ability to reclaim our communities is about us creating a future for ourselves so that we don't feel like we've got to move out of our neighborhood and live someplace better. I think that is the crux of the work that environmental justice is supposed to be addressing, but I'm not sure if it always does because it seems like many folks within that realm are stuck highlighting the wrongs that have been perpetuated against us. I'm not saying they're wrong for hammering that home, but I also think that the other part has to be creating the future that we want, and that's what I feel like I've been doing for the past 20-some years.

LEARN MORE

majoracartergroup.com/about

Reclaiming Your Community: You Don't Have to Move Out of Your Neighborhood to Live in a Better One, Berrett-Koehler, 2022

> I've known Majora for many years. We both understand the importance of green space in urban places, and the importance of building community by providing spaces for people to gather and exchange ideas. In order for a community to thrive, it needs be stable and its people healthy and free of industrial waste. Majora has dedicated much of her life to helping others embrace that philosophy. —DUDLEY EDMONDSON

866B

THE BOOGIE DOWN GRIND

THE BOOGIE DOWN GRIND CAFÉ

Powered by StartUp Box

Support

MAJORA OUTSIDE HER BOOGIE DOWN GRIND CAFÉ.
COURTESY OF MAJORA CARTER

CORINA NEWSOME

ORNITHOLOGIST

ATLANTA, GA

Corina Newsome has always been candid and a truth-teller, regardless of how people feel about it. She has thought-provoking things to say, but she also has this infectious joy about birds. Birds bring her joy; when she sees a beautiful bird, her excitement, like that of a kid in a candy shop, makes you excited too. Her joy is a great thing to experience.

—DUDLEY EDMONDSON

CORINA BIRDING IN DOWNTOWN ATLANTA
BY DUDLEY EDMONDSON

WILDLIFE CONSERVATION AND HUMAN CARE

I am a wildlife biologist with a passion for avian conservation. My graduate work focused on nest predation in a small coastal bird called the seaside sparrow, to better understand the ecological dynamics at play in salt marshes along Georgia's coast. Even as a wildlife scientist, my work is rooted in the advancement of equity and environmental justice for human communities in conservation science, policy, and practice.

CORINA STANDING BETWEEN HER MOTHER AND FATHER WITH HER SIBLING IN THE STROLLER
COURTESY OF CORINA NEWSOME

MY PARENTS INSPIRED MY LOVE OF NATURE

I grew up in Philadelphia in an urban setting where there weren't a lot of easily accessible natural spaces. So, oftentimes, my mom would put us on a bus, and we'd go out into the suburbs. She would take us into natural areas every opportunity she had, because even though we lived in an urban environment, she loved connecting with nature. My passion for wildlife was ignited by the many books, encyclopedias, and magazines my parents and grandparents provided. Before my dad passed when I was young, he had collected shelves and shelves of books about nature. I would read them over and over and over again, fanning the flames of what would one day become my career. Though I didn't realize my passion for birds until adulthood, I distinctly remember my excitement about American robins. We had flocks of them that would come through our neighborhood in the park, down the street. That park was maybe an acre of mowed grass with a willow tree in the middle. Every couple of years, we'd have a flock of robins land there, and my mom would say, "The robins are back!," and we'd all run outside and down the street to look at the robins in the field.

BIRDS ARE FLYING, WALKING, CHIRPING BODIES OF HOPE

Birds bring me a lot of enjoyment. I love hearing them, seeing them, and witnessing their adaptations. They embody hope for me. One of the reasons why I love the poem by Emily Dickinson, "'Hope' is the thing with feathers," is because I feel hope in my bones when I think about birds. Their existence seems impossible, yet still they thrive. I'm constantly in a state of joyful anticipation. It's winter right now, and I cannot wait for the first hummingbird to return. I cannot wait until I hear my first summer tanager echoing in the forests. There is no joy quite like bird joy.

THE FUTURE OF ENVIRONMENTAL JUSTICE IS LOCAL

When it comes to environmental justice, there are solutions that can bring community restoration, and those that can exacerbate already-existing harms. If the communities facing environmental burdens are not positioned to have the resources and decision-making power for place-based environmental problem-solving, then they will continue to be exploited, even in solutions to environmental challenges. I want to see leaders equipped to lead in their own communities. If you are equipping community leaders, you need to understand you have nothing to offer them regarding how to make information relevant to their community. All you have is the power and the money. You need to give your power and money to people who are from communities that have been disinvested in, who have ecologically barren spaces because of your exploitation. You need to remove yourself from the position of decision-maker in every possible way.

You need to give your power and money to people who are from communities that have been disinvested in, who have ecologically barren spaces because of your exploitation.

It is critical that conservationists take an intersectional approach to the way that they prioritize where they invest time and money, realizing that they have to pay attention to the environmental vulnerabilities of humans and wildlife. This includes urban spaces. It includes finding ways to make even power line right-of-ways and lots and parks that may have fallen into disrepair more beneficial for people and wildlife. And it must be done with care and intentionality. Part of this intersectional approach may include working with municipal leadership, such as in housing and community development, to make sure that legacy residents aren't displaced by green-space investments.

There is research by professor and community organizer, Dr. Na'Taki Osborne Jelks of Spelman College in Atlanta, Georgia, that details the problem of green gentrification—the process by which economic investment in green spaces leads to displacement and destabilization of the legacy, Black and Brown low-income communities around those green spaces. There needs to be interdisciplinary and community-centered collaboration to ensure that the people who need the green spaces are the ones who benefit from the investments. Caring for the ecosystems where people live, especially communities bearing the heaviest weights of environmental harm, is one of the most meaningful wins that we can have in wildlife conservation.

PARADISE LOST

In February 2020, Ahmaud Arbery was murdered. Because of racism and corruption in local leadership, it didn't hit the news until late spring. I remember reading the *New York Times* article that broke the news nationally, and then going into the field the next day, passing the neighborhood where he was murdered on the way. I was sickened by the fact that Ahmaud and his beautiful family were ripped apart by hate and violence, when they should have had

the freedom to thrive in Georgia's beautiful coastal ecosystems.

I was grieved by the fact that his life was stolen, and nobody cared enough to do something. When I say nobody, I'm talking about white people. They saw it happen and tried to protect the murderers. Then, I felt guilty. Black people have to fear for their lives in this city, in this neighborhood, and I was lollygagging in marshes, looking for seaside sparrow nests.

I didn't deserve to do this when other Black people are being murdered, and no one is being held accountable. My entire perspective on the science that I was doing was distorted because racial violence was allowed to extinguish Black life in these "golden isles." The place and science that I loved became like a curse. I repeatedly thought to myself, "I can't spend my time like this. I should be doing something more important that benefits and protects my people."

RACE AND CLASS ARE BIG FACTORS IN DETERMINING ACCESS IN CONSERVATION AND STEM CAREERS

After graduating from college, the first leg of my career was spent as a zookeeper. From the time I became interested in conservation as a

A NORTHERN MOCKINGBIRD, ONE OF CORINA'S FAVORITE BIRDS, SINGING FROM A BRANCH
BY STAN TEKIELA

child until my first professional experience, I had only ever seen white wildlife scientists. It was not until my first internship at the Philadelphia Zoo, at the age of 17, that I met a biologist who looked like me. Her name is Michelle Jamison, and she was the lead carnivore keeper at the zoo, specializing in the breeding and care of imperiled species. I was stunned that someone like her existed.

But if I had any questions about the whiteness of conservation professions after starting that first internship, I soon had answers. In order to become an animal care professional, you had to complete several years of unpaid internships. Wealth, like many other forms of security, falls strongly along racial lines in the US. Requiring years of free labor from young pre-

professionals has disproportionately excluded Black and Brown people from the world of conservation, and the proof is in the whiteness of the workforce pudding.

For years in college, and even after becoming a zookeeper, I advocated for internships to be paid to break this cycle of exclusion, only for my "colorblind," and often boldly racist, white colleagues to tell me we had to "pay our dues." It was not until I met international conservation and education leader Dr. Brian Davis that I saw someone with power changing course. Dr. Davis was the president and CEO of Georgia Aquarium, one of the highest-ranked aquariums in the world, and the first Black chair of the board for the Association of Zoos and Aquariums. Among his many significant contributions, he used his power and influence to hold the conservation community accountable for paying interns for their work, addressing a significant barrier to the participation of aspiring conservationists from Black, Brown, and low-wealth communities.

In addition to making systemic changes in the conservation field, Dr. Davis took the time to mentor countless young Black and Brown conservation professionals, understanding that the representation and insight he provided was rare and incredibly important for those of us beginning to navigate this field. My hope is that I can model Dr. Brian Davis's commitment to advancing equity in conservation, and work toward a future where a young Black child who loves wildlife is not shocked to see a Black biologist—they expect to see us.

LEARN MORE

#BlackBirdersWeek

#BlackinNature

#RepresentationMatters

> When I first approached my publisher about the book, one of the editors said "Oh, you've got to get Corina Newsome for this book!" As a fellow birder, I already knew Corina and agreed she was a great choice. She speaks from the heart about how racism and colonialism affect scientific study and nearly every aspect of our society. Her passion for birds and BIPOC in science makes me proud of her and her generation. —DUDLEY EDMONDSON

THE PHILADELPHIA MUSEUM OF ART, WHERE CORINA'S MOTHER TOOK HER AND HER SIBLINGS; BECAUSE HER OWN NEIGHBORHOOD LACKED GREEN SPACES, IT WAS HERE THAT CORINA LEARNED MUCH ABOUT THE NATURAL WORLD. ACESHOT1/SHUTTERSTOCK

ROXXANNE O'BRIEN

ENVIRONMENTAL JUSTICE ACTIVIST

MINNEAPOLIS, MN

Roxxanne O'Brien stands out for her direct involvement in environmental activism, notably using her voice and presence on the front lines against pollution sources. Recognized for effective community organizing and her ability to mobilize for clean air rights, her efforts with the EPA to improve local industrial practices exemplify her commitment. Our conversation highlights the importance of building trust, allowing Roxxanne editorial control over her contributions to ensure authenticity and respect, contrasting with experiences of exploitation. Her proactive approach, supported by her community, focuses on achieving justice and meaningful change.

—DUDLEY EDMONDSON

ROXXANNE, FRONT AND CENTER, STANDS WITH HER FELLOW ENVIRONMENTAL JUSTICE ACTIVISTS, INCLUDING PRINCESS HALEY (FIST RAISED), ROSE YOUNGMARK (HOLDING A MEGAPHONE), AND DANIELLE SWIFT AND HER CHILD.
BY DUDLEY EDMONDSON

ALWAYS A COMMUNITY ORGANIZER

I live in North Minneapolis, which is made up of 13 neighborhoods. I've lived in 5 of the 13 neighborhoods, and I currently live in Folwell. I started as a trained community organizer working on equity issues within the park system. I've also worked on issues around housing, gun violence, food justice, electoral politics, and police brutality. I'm mostly known as an environmental justice organizer. I've been on many committees, councils, and coalitions for organizations and even the city of Minneapolis. I still consider myself

ROXXANNE POSES WITH HER MOM FOR A SELFIE
COURTESY OF ROXXANNE O'BRIEN

a community organizer because I don't work on these issues in silos since they're all connected. Organizers try to connect issues, people, money, and power to create change in the world. I don't see any other life to live, outside of working on issues that impact me and my loved ones.

AS A CHILD, I LEARNED A LOT FROM MY MOM AND WATCHING HER BATTLE RACISM

I was born in L.A. My mom and I moved to Minneapolis in 1987, when I was about 4 years old. She moved us here for more opportunities in response to the increased need for teachers of color in the North. When we got here, I started to smell and recognize poison in the air for the first time. As a child, I could not have known there was a whole industrial corridor between North and Northeast Minneapolis that was responsible for the toxic odor. I remember riding in a car and asking my mother "What's that smell?" . . . and she didn't know. I would always wonder because I smelled it all the time.

One of the things that stood out to me as a kid was watching my mother battle racism and seeing her come home crying. As a Black teacher in the school system, she had to work with white people who

A GROUP OF PROTESTERS DEMAND JUSTICE AT A NORTHERN METALS FACILITY.
BY ADAM REINHARDT

were anti-Black and especially anti-smart-Black woman. My mom's pretty strong, but they did everything they could to undermine her and break her spirit. I can count on one hand how many times I've seen her cry. Those times were shocking and stuck with me. I knew then that there were some serious issues with white people.

I learned a lot watching my mother. She was my first teacher. She knew a lot of history. She also has multiple master's degrees. I got a head start as a child, learning about community issues before I understood them. I remember being with my mom everywhere, at radio stations, community meetings, and other events. I would be under the table, listening to grown-ups talk about issues in our community. I watched her work the room, support parents and families, and build relationships. That was my earliest training in community organizing.

ENVIRONMENTAL JUSTICE FOR MY COMMUNITY

After I learned some community organizing tools and skills, I started to build relationships in the community. At one time, I was vice chair of the North Minneapolis chapter of the National

Community Reinvestment Coalition. I did foreclosure prevention work, trying to save wealth in the community. I worked with elders and Black and Latino people, the people targeted by predatory practices the most. After doing that work, people started to notice that I would speak up about things. It came in handy because it meant people would send me information. That's how I learned about north Minneapolis's Hawthorne neighborhood.

I've heard, if you want to know how much corruption is in your community, pay attention to how much pollution is in your community.

LEAD IN HAWTHORNE

The Hawthorne neighborhood not only had the highest foreclosure rate in our community, but it had the highest lead levels in the state as well. I've heard, if you want to know how much corruption is in your community, pay attention to how much pollution is in your community. The more pollution, the more corruption. There are a lot of people who used to work as politicians or public officials who are now lobbyists for polluting corporations. Someone forwarded me information about a facility called Northern Metals, and it said they wanted to increase their pollution level. This would mean anyone

in the Hawthorne neighborhood trying to grow a vegetable garden would likely have a much greater chance of getting cancer than in most neighborhoods in Minneapolis. The permit increase request would not say the specific amount outright; it would say something like 0.85, but no one in our communities knows what that means. We don't know the language. We don't know what particulate matter means. We're never really made aware of these types of issues in our community or even involved in the conversation, even though they impact us more than other people. Once I got that information about Northern Metals, I started telling people everywhere I went. The narrative told about our communities is that we're unhealthy, we're violent, there's Black-on-Black crime. Those are the narratives about poor communities or, as I like to say, exploited communities, because the word exploitation tells the truth about what's happening.

THE FIGHT AGAINST CORPORATE CORRUPTION

Northern Metals didn't want to admit how much they were polluting. They were sloppy culprits, and they were bullies. They have a bunch of lawyers and any time the city or the community would stand

PILES OF SCRAP METAL BEHIND A FENCED-IN AREA IN A NEIGHBORHOOD NEAR DOWNTOWN MINNEAPOLIS
BY DUDLEY EDMONDSON

ALLOW GATE TO
FULLY CLOSE
FOR SECURITY
REASONS

TWO PROTESTERS OUTSIDE THE NORTHERN METALS RECYCLING PLANT
BY ADAM REINHARDT

up, a court proceeding would be involved. The state's attorney general got involved because Northern Metals was trying to use legal tricks to rush the approval of their permit by the Minnesota Pollution Control Agency. There was a lot more pressure to raise the alarm about them. Public officials and people started to pay attention and also joined in. Eventually, it came to a point where the EPA, I think, had put an administrative order on Northern Metals to stop their shredder usage. That's when you started to see a big shift in the direction of where Northern

Metals was heading. After that, there was private mediation, but every chance I got, I still kept the community's demands at the table. I thought, if there's any money, it should go back to the community. It normally would have gone back into the state's environmental fund, but we were able to get $600,000 back into the community for asthma programs and education and any sort of respiratory support for residents. That was one of the historic wins that we had. We still have some of that money. And it's been work holding the city accountable on how those funds get spent. The work never really ends, having to sound the alarm so many times to get people to do their job.

MY CHILDREN, MY LEGACY

I believe in the power of people and working to do my part and pay my rent to this earth. One day I had an experience with life that changed my outlook. I decided I didn't want to play small anymore. I realized how valuable life was, and I didn't want to be taking up space on the earth. I want to leave something, add something, so I've been pretty determined ever since. I regret the times I've missed with my children to carry out this work. When you know what you know, you have to do something with that knowledge. I feel accountable to my ancestors, and responsible to my community, because these issues impact us all. Just because others pretend to not see these issues and the connections, doesn't mean I should too. It's a choice I make every day. I like to move with as much integrity as possible and to move toward the fear, even if I'm afraid. I've read some books by Nelson Mandela where he questioned whether everything he did was worth it, considering what happened with his family. I constantly think about that, but I also see the victories, and I see what's happening and how it impacts my community and my family. I think, as a parent, it's my job to try to do better. I hope that I get a chance to breathe and relax with my children at some point. I hope I'm able to leave something for them.

LEARN MORE

National Community Reinvestment Coalition
ncrc.org

I first came across Roxxanne in a photo—she was standing outside a factory carrying a bullhorn. That changed everything. I knew she would be a strong, no-holding-her-tongue voice who could speak to the issues of industrial waste contamination within low-income and Black communities. It's really been an honor to work with her, and I'm thrilled she said yes to being part of this book project. —DUDLEY EDMONDSON

POLLUTION FROM THE UNITED STATES IS A MAJOR DRIVER OF CLIMATE CHANGE.
BY AYDO8/SHUTTERSTOCK

WHAT?!

"The United States is the
world's second-largest polluting
country, pumping more than
6 million metric tons of CO_2
into the air each year."

—ENVIRONMENTAL PROTECTION AGENCY

DEJA PERKINS

GEOSPATIAL ANALYST AND URBAN ECOLOGIST

RALEIGH, NC

Deja Perkins's research high-lights the importance of urban green spaces, which provide a direct connection to nature and enable residents to become better stewards of the land. Too often, wilderness areas—where few people live—are favored, but urban green spaces are important not only for wildlife, but for mental and physical health of city residents.

—DUDLEY EDMONDSON

DEJA AT THE SARAH P. DUKE GARDENS IN DURHAM, NORTH CAROLINA
BY DUDLEY EDMONDSON

BRINGING SCIENCE TO PEOPLE

I received my bachelor of science in environmental science, natural resources, and plant sciences with a concentration in wildlife from Tuskegee University. At one point, I wanted to become a wildlife biologist, but during my time at Mississippi State University, I learned that a lot of that work would focus on pristine, wild, and untouched places. Originally from Chicago, I knew I wanted to live near or in a city, so I traveled to North Carolina to pursue my master of science in fisheries, wildlife, and conservation biology at NC State University. During that time, I studied bird diversity across the Triangle and found that a popular bird recording app lacks data in predominantly low-income and majority BIPOC areas, even though birds exist in those areas. Now I investigate the consequences of a lack of inclusion in participatory science and how we can make projects more inclusive, diverse, equitable, accessible, and locally relevant.

GROWING UP IN CHICAGO, NATURE WAS A FANTASYLAND TO ME

My family migrated from Alabama during the Great Migration and now my whole family lives in Chicago. I grew up in the heart of Chicago on the Southeast Side, between Lake Michigan and Indiana. I lived a sheltered life growing up, because Chicago can be dangerous. Lake Michigan was between my house and downtown, so I have fond memories of seeing Canadian Geese out of the car window grazing on the Museum of Science and Industry lawn, the beautiful trees in Hyde Park, and the seemingly endless shores of Lake Michigan. As we drove, I would see nature transition from very few trees around our neighborhood to many trees in the Hyde

DEJA, IN THE SIXTH GRADE, STANDS NEXT TO A HISTORY FAIR ENTRY.
COURTESY OF DEJA PERKINS

Park area, then few trees again as we entered downtown. I would see the opposite transition as we left the majority-gray urban center and drove to the suburbs. There was not only a rapid change in the environment but a rapid change in the economics of the neighborhoods. That always made me curious, and I think it's the foundation of why I study inequities of access today.

As a kid, my mom kept me at the zoo, museum, and aquarium, thanks to free passes from the library. Those are my fondest memories. I had a special interest in animals growing up. Those places were my strongest connection to nature because I didn't really see nature outside my window. For me, nature was inside of a book. It was a fantasyland. It was something that I saw on TV, on Animal Planet, and National Geographic. They'd show these faraway places like Thailand, the savanna of Africa, and the Antarctic. At the time, those destinations weren't places that I could imagine myself going, but being able to go to the Shedd Aquarium, the zoo, those were real connections for me.

When I was growing up, there weren't a lot of programs where I could be exposed to nature and the local ecology of where I lived. The only local program I could find was a small family organization called Fishing Buddies Inc. Their goal was to get inner-city kids introduced to the outdoors and the local nature that surrounded them. They took us on field trips to different nature centers and forest preserves that were both near and far from our neighborhood. That experience with Fishing Buddies was my first exposure to nature and the outdoors, but I didn't really start to explore the outdoors until undergrad and grad school.

NO ONE TAUGHT US ABOUT THE INDUSTRIAL WASTE AROUND US IN SCHOOL

Some of the urban green spaces I visited through local programs were 10 minutes from my house, but also right across the street from a landfill. I study environmental justice now, but I didn't realize I grew up surrounded by issues of it. It was a part of my everyday experience and I had no idea. They weren't teaching us about the impact of the nearby landfills in school. That wasn't in our curriculum. Growing up, I remember my mom complaining about how bad the smell was, but since I was born without a sense of smell, I never realized how problematic that could be. The landfill and factories were a part of our everyday existence, and we knew to automatically roll the car windows up. Was being born without a sense of smell an impact of growing up around that area? All my siblings have asthma. Was that from growing up there? All those health connections could be tied to the landfills, the industrial waste plant, and the other plants and factories that were right across from the green spaces.

SAFETY IN THE OUTDOORS

Birds are what got me into the research that I study now. I knew I wanted my work to focus more on urban environments because that's where I grew up. A researcher at NC State was organizing a local bird count of the Triangle region of North Carolina. That project was great because it allowed me to go birding in regular areas, like neighborhoods. It wasn't all parks and green spaces. I was birding in front of people's homes, walking up and down streets. I was getting to explore Raleigh, and I got to see how the economic gradient would change as I drove through different neighborhoods. I also noticed a difference in the types of birds that I would see.

That was the first time I noticed that the bird species may differ in low-income areas vs high-income areas. That was also when I was

introduced to the idea of safety in the outdoors. Unfortunately, I didn't feel safe in either low- or high-income neighborhoods. It didn't matter if it was a wealthy, predominantly white neighborhood where people were generally a bit nicer because people still felt comfortable walking up to me, asking, "Oh, what are you doing?" One time, I had a lady follow me with her dog in one of the newer developments. She told me she didn't want me birding in front of her home because she thought I was spying on her with binoculars. She didn't feel comfortable with me being there even after I told her what I was doing and showcased my official T-shirt and paperwork.

When I would go into areas where I felt like my Blackness would protect me, I would get cat-called by people driving by, and there were a few instances where people would follow me in their car and circle the block. In some places, I felt like my Blackness was perceived as a threat, and in others, I felt like my gender made me a target. It was a harsh awakening.

ENVIRONMENTAL DISPARITIES BETWEEN HIGH- AND LOW-INCOME NEIGHBORHOODS

I learned a lot from working with the Triangle Bird Count. Seeing

DEJA POINTING TO A DATA MAP SHE CREATED; TAKEN IN HER OFFICE ON THE MAIN CAMPUS AT NORTH CAROLINA STATE
BY GEORGE HARLEY-WENN

the differences in vegetation, tree structure, and bird diversity across different neighborhoods in the Triangle region of North Carolina was eye-opening and confirmed my childhood hunches. But when you look at the data, the important part of this whole project was the difference between the systematic count that I was conducting as a technician with the Triangle Bird Count and the randomness of observations that were submitted through eBird. eBird is an app and website hosted by the Cornell Lab of Ornithology that helps birders keep track of their bird observations, but as I examined the data as a scientist, I found that it was also showing a reflection of its users and who was reporting the data.

When I mapped the data and overlapped it with other data sources like median household income, I could see that a lot of the points from eBird are in high-income areas. There was barely any data at all within dense urban areas. It made me question why people aren't going birding in these areas. A recent survey study published by a colleague showed that 96% of eBird's users are white, high earning, and educated. I found this is reflected in the data because majority-white-occupied and high-income areas are being sampled. The lack of data in high-minority census tracts and low-income census tracts shows that there is a gap in who's reporting the data. When we lack that information and data, we're completely erasing the stories of communities and the types of birds and biodiversity they have access to.

That's exactly why diversity matters, including diversity of experiences and of thought. Academia and many other sectors of society have lost sight of that. Scientists are in service to society. We are supposed to research the topics that are important to the people. If you don't have a variety of people and lived experiences represented, then your research is only catering to a certain audience, and that is a disservice to the rest of society.

> **If you don't have a variety of people and lived experiences represented, then your research is only catering to a certain audience, and that is a disservice to the rest of society.**

PEOPLE-POWERED DATA CAN GIVE THOSE UNDERSERVED BY SCIENCE AGENCY

eBird does all this fancy statistical analysis to summarize, track, and predict where birds are going to be, and that's super helpful. However, it doesn't help when people are trying to discover what the local biodiversity looks like in their neighborhood. eBird encourages users to explore "hotspots," or areas with confirmed high numbers and species of birds, but these tend to be those locations where the habitat is more pristine or in parks in higher-income and mainly white-occupied areas. However, if more diverse birders explore and report around their neighborhoods, we can have a better understanding of the spaces that birds and people use across the city. That's another core piece of why I like using people-powered data. It has the potential to give people agency and a voice in science through something as small as collecting or reporting data.

However, many people, especially those traditionally underserved by science, aren't aware of participatory science or its ability to give us agency in the spaces where we live. I think that highlights the importance of access. If people don't have exposure to opportunities, they won't be aware they exist, and won't have the access. Without access, people don't have the opportunity to make nature their safe space, form connections to care about the environment, or allow nature to help with their mental health.

BIRDS HELP ME WITH MY MENTAL AND PHYSICAL HEALTH

After my master's work, I decided to pivot to a more computer-based research because of health issues and having to navigate science as a person with multiple disabilities.

DEJA'S MOTHER AND GRANDMOTHER BIRDING TOGETHER
COURTESY OF DEJA PERKINS

I have a condition called syncope, where I sometimes have a temporary loss of consciousness caused by different triggers, like heat, pain, or standing too fast. I also have occipital neuralgia and chronic migraine disease, which makes me sick for about 15 days a month. I have been exploring the process of being tested on the neurodiversity spectrum. I want to be a part of breaking generational curses and understanding why certain things are so difficult for me and what strategies I can implement to be happier or make my life easier. It has been a very interesting process because there's a lot of negative stigma around mental health, neurodiversity, seeking help, and doing therapy. I think being open with my challenges is another way to break down barriers. I haven't necessarily been courageous enough to talk about it openly. There have been a variety of things going on in my life that make pursuing this academic route difficult.

Birding is one of the ways that I have tried to incorporate happiness back into my life. I incorporate nature into my mental health practices because it's sort of like an escape, but it's also a relief, and a way of being able to get out of my own head. I'm sure we all have our own struggles with anxiety, imposter syndrome, or never feeling like we're good enough. The pressure of success can be harmful.

I use nature as my way to disconnect from the outside world and all the things that society tells us that we should be doing or have and use it as a way to connect back with myself and my ancestors. Nature makes me remember that the world is so much bigger than myself, and I think everyone needs to be reminded to think beyond themselves. That is also another driver of the work that I do. I want to help people in my community and make sure that change happens in places like where I grew up, so that people can have more opportunities to be more informed and even potentially help make decisions and push for change in their neighborhood.

LEARN MORE

Fishin' Buddies
fishin-buddies.net

Shedd Aquarium
sheddaquarium.org

Lincoln Park Zoo
lpzoo.org

Brookfield Zoo
czs.org/Brookfield-ZOO/Home.aspx

> Deja is another bright, young Black woman in science. The research work she does around birds is shedding light on some of the racial biases within science and the birding community. Her work, and our conversation, really got me thinking more and more about the importance of urban green spaces to the overall environment and the people who live near them.
>
> —DUDLEY EDMONDSON

DEJA LEADING A BIRDING HIKE AT THE SARAH P. DUKE GARDENS IN DURHAM, NORTH CAROLINA
BY DUDLEY EDMONDSON

QUETA GONZÁLEZ

DIRECTOR OF CENTER FOR DIVERSITY & THE ENVIRONMENT

PORTLAND, OR

Queta González's work encourages folks to expand their understanding of systems that benefit some over others and offers them the tools they need to be better community members, which makes them better equipped to engage across differences. She makes it clear that there's cause for hope and determination—understanding we likely have more in common than we don't. Bringing people to your cause, regardless of race, is one way to ensure that the things you're doing can continue, long after.

—DUDLEY EDMONDSON

QUETA SITS ON A PARK BENCH IN PORTLAND, OREGON.
BY DUDLEY EDMONDSON

AN ARC TOWARD COLLECTIVE LIBERATION AND LOVE

I've been with Center for Diversity & the Environment (CDE) for 14 years. The work of the Center for Diversity & the Environment is always evolving in an arc toward collective liberation and love. We are a small, mighty, and scrappy organization. The team works together in running the organization, developing and delivering programs, and staying involved with community. The daily work of culture change can be challenging, so we are intentional about 'joy spotting.' It's important to pay attention to the beauty that

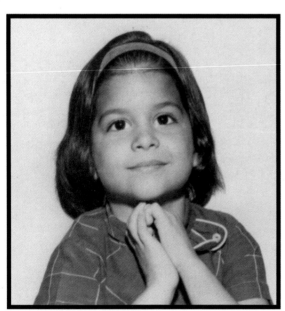

QUETA AS A YOUNG GIRL, APPROXIMATELY THE AGE SHE ATTENDED SCHOOL IN SIOUX FALLS, SOUTH DAKOTA
COURTESY OF QUETA GONZÁLEZ

surrounds us because it is a source of clarity, connection, and energy.

AS KID, BEING OUTDOORS WAS ALWAYS A PART OF LIFE

I'm originally from Venezuela, and my early years were spent primarily in Caracas. Just to give you an idea of what life was like, our home had an open patio in the middle, so the middle of the house was outdoors. There was a garden, there were chickens, a turkey, a guava tree—there were all kinds of plants and critters. My mom is from the United States, and my father from Venezuela. They met in Kenosha, Wisconsin, when they were going to university.

My mom, brothers, and I came to the United States initially because my mom's father was dying. We came to meet our US family. Our parents wanted us to have some of our education in the United States, then return to Venezuela. That would bolster our cultural competence and strengthen our English. Eventually, my parents grew apart and got divorced. My mom raised us as a single divorced mother in South Dakota—in the 1960s. Single parenting is difficult enough, but in those years being divorced was frowned upon. Despite a less-than-gracious community attitude, my mom saw good all around. My mom

QUETA RAFTING THE COLORADO RIVER IN THE GRAND CANYON, 1986
COURTESY OF QUETA GONZÁLEZ

grew up on dairy farms—farming and gardening—and drew on that as she raised us. We ate whole foods and practiced living a subsistence life. We grew our own vegetables and canned them; we fished, we hunted. Being outdoors (and chores) was always a part of life.

AS A SCHOOLKID, EVERY TIME SPANISH CAME OUT OF MY MOUTH, THE TEACHER WOULD PUT ME IN THE JANITOR'S CLOSET

I lived in Sioux Falls, a predominately white town of about 80,000 at that time. I understood English because my mom and dad spoke English and Spanish, so, for a time, I spoke both. In fourth grade, I had

this teacher who was wretched, and every time Spanish came out of my mouth, she put me in the janitor's closet. I never told my mom about it because, in my 9-year-old innocence, I believed I must have done something terrible.

That year was followed by another teacher who started our lesson of studying South America with Venezuela. My home! I was so excited to share this part of myself. Her opening line was, "We're going to start with Venezuela, the northernmost country in South America, where people are backward, live in tin can huts with tar paper roofs, and are completely uneducated." I stood up and I said, "No, you don't know what you're

QUETA AND HER MOTHER, SONYA, SITTING TOGETHER
COURTESY OF QUETA GONZÁLEZ

talking about." She sent me to the principal's office. The secretary wasn't in the office, so I walked out the front door, turned right, and walked home. When I got home, the phone was ringing and my mom said, "Queta, what's going on? Stay put. I'm on my way home."

When she got home, we walked back to school together as I told her the story of what happened. I was sitting outside the principal's office, while my mom went to talk with the principal. I could see her in the office and her hands were flying around everywhere as she spoke. Once again, I was sure I'd done something wrong. I thought, "My gosh, I'm in trouble." I was called into the principal's office. When I walked in, the principal apologized to me. Then Miss Smith came in, and she apologized to me as well.

What I learned that day was that my voice mattered. I learned that my mom would always stand up for me and have my back. I understood that my difference mattered and it's the people in our lives that influence the direction of our lives.

Heading into sixth grade, there was a wonderful teacher—thank goodness. I caught a break! She was one of the first people I ever met that used the "Ms." salutation. Ms. Borich was phenomenal. She read *The Hobbit* and *The Monkey Wrench Gang*, a book by Edward Abbey, aloud to our entire class. That book really caught my attention. I started thinking about what we are doing to the environment, and I haven't stopped thinking about it since. Again, it's those people lifting you up. When the world is telling you you're wrong, you're not good enough, and coming down on you in all sorts of ways, it's people like Ms. Borich who help, people who were integral to me finding my way, and I owe them. If I didn't have them, I would have been lost in a lot of ways.

RETHINKING EVERYTHING TO ACHIEVE COLLECTIVE LIBERATION

I think Center for Diversity & the Environment has a unique founding story. About 16 years ago, a group of folks started the Environmental Professionals of Color network. They received funding to hold forums within the city of Portland and beyond. The group convened people from traditional environmental and conservation organizations, as well as social and environmental justice backgrounds. The conversation focused on what was happening in the environmental movement with regards to, inclusion, and equity.

One of the requests that surfaced was that an organization be formed that specifically works with people of color in environmental and conservation fields to evolve and transform the movement into a more inclusive, diverse, and equitable place. So, with that, the community essentially co-founded CDE.

At CDE, we meet people where they are, and work to co-create a space for learning and unlearning. We examine systems that we live in, the systems that perpetuate a dominant culture of white advantage and capitalism. Those systems feed other systems that degrade our planet, our relationships, and our ability to see each other within the idea of kinship. We encourage

It's critical that people can envision something different because if you can envision it, you can create it.

people through self-reflection and deeper analysis, to move into action. It's critical that people can envision something different because if you can envision it, you can create it. If we can get enough people to envision a new way of being and doing, then they can move toward collective liberation and healing. We can begin to see healing in our relationships, not just with humans, but with all of our relations. Everything is interconnected, and our fates are linked together.

Our work focuses on supporting the learning environment and classrooms of program participants to equip them to evolve their personal and organizational culture and analysis to be more inclusive, diverse, equitable, and just. Our focus is on moving from awareness to deepening of analysis, then integrating learning into thoughtful action. And ultimately, reflection and adjustments that incorporate the learnings brought about by those actions. The work that we do is centered on healing, kinship, love, and liberation. We believe that change is brought about when we engage change on the individual, interpersonal, organizational, systemic, and movement levels.

The program participants have got to see themselves in it. They've got to see the impact of structural racism on their own lives in order to have consistent motivation. They have to see how those systems play out in their relationships and in their organization and move to change it. That takes a real intentionality and clarity on why it's important. Whenever people look at this as an add-on, something they say, "We really think it's important, but we don't have time." That indicates that they're still separating themselves from the reality of their own benefits from these systems or they lack the understanding of how the systems are keeping them from being inclusive.

The author Gloria Anzaldúa once said, "I change myself, I change the world." When we can start identifying how our implicit associations, how dominant culture comes through in our actions, behaviors, and values, we can then start unraveling that and move to this place where we are really building authentic relationships across differences and valuing those differences. Then we can take a look at systems and start re-imagining them and move toward liberation.

QUETA AT CITY OF ROCKS NATIONAL RESERVE IN SOUTH-CENTRAL IDAHO
COURTESY OF QUETA GONZÁLEZ

LEARN MORE

Center for Diversity & the Environment
cdeinspires.org

Queta is a big-picture kind of person. Her work at the Center for Diversity & the Environment centers on creating a diverse set of changemakers who will chart a new course for the environmental movement. The hope is that, by doing so, we will all be in better relationship with each other and the natural world we depend on for our existence. —DUDLEY EDMONDSON

DR. DREW LANHAM

ECOLOGIST AND BIOLOGIST

CLEMSON, SC

Drew Lanham knew of me before I knew him. He saw me on a Nickelodeon program where I was introducing some kids to birds and bird-watching. Drew saw that and said it really affirmed his feeling of a love for birds, as he'd never seen another Black man who was a birder. Drew's work carries that on, and makes it clear that what BIPOC environmentalists and birders are doing isn't something strange or crazy.

—DUDLEY EDMONDSON

DREW SITS IN HIS FAVORITE WRITING SPACE, SUNSET CAMP,
A PIECE OF PROPERTY HE OWNS OVERLOOKING A LAKE.
BY DUDLEY EDMONDSON

IDENTIFYING WITH NATURE

I'm a conservation ecologist, and I teach at Clemson University. My fieldwork really began with trying to understand how forest management practices like clear-cutting, for example, impact different bird species. I also call myself a cultural and conservation ornithologist. I do that through birds. I see birds as messengers and emblems of our lives, as many cultures have done for tens of thousands of years. I try to bring all that together through the work that I do as a scientist, writer, poet, and speaker, hopefully engaging people empathetically.

DREW AS A SMALL CHILD, AROUND THE TIME HE LIVED WITH HIS GRANDMOTHER
COURTESY OF DREW LANHAM

Ultimately, my job is to try to get people into a mind space where they're not just thinking about identifying that warbler by wing bars or songs, but also identifying with that warbler because it might be declining or identifying with a habitat because that habitat is home not just for the bird, but for them too. What I do as a university professor is try to expand how people think.

A CHILDHOOD LOVE AFFAIR WITH BIRDS

It all started in the boondocks of South Carolina. I grew up in a place with dirt roads, no sidewalks, no streetlights. All of our food came from the garden or from the pastures where we raised cows. Occasionally, my dad would kill a rabbit, and we'd have it for supper. My earliest bird education came from my grandmother's ornithology, listening to the names that she had for birds before I learned the names out of field guides. The inspiration for my love affair with birds came from wanting to fly. I wanted to be a pilot for a long time, so I could fly, but then birds could do things that pilots couldn't do. Birds were disappearing to foreign places that I never had any inkling or any imagining that I would be able to go. I lived vicariously through

A SCARLET TANAGER PERCHES ON A BRANCH IN THE FOREST.
BY DUDLEY EDMONDSON

birds, watching them fly. My idea was that, somehow, I could replicate what birds did. I built wings out of cardboard and tried floating on umbrellas from barn roofs like Mary Poppins. I did those things to try to be a bird, but learned pretty quickly that gravity was always going to win. My goal became to learn as much as I could about birds. When I saw a scarlet tanager for the first time, I was amazed at the brilliance of it and how it came from the same place where jaguars and harpy eagles lived. The world opened up to me through birds.

CONSERVATION IS A MORAL ACT

Conservation to me is just another word for caring, but caring in a way that saves something for somebody else. Conservation is a moral act, absent any god. Morality doesn't require a god. Morality requires us to think about the well-being of others, not just the well-being of non-human others, but the well-being of other human beings. It doesn't make sense for someone to say they love life and then condition that on, well, it has to be four-footed or feathered or finned or white-skinned, otherwise, I hate you. If you tell me you care about nature, then care about all of nature. Don't pick and choose the birds or the beasts that you want to love. Because then you got me in a shopping cart in the discount line and you're discounting me ahead of everything else, but you expect me to jump on your bandwagon and be with you because you want to save the world. Really, what you're telling me is, "I want you to help me save the world for me, and you can have what's left in your marginalized corner."

This is part of the problem with American conservation. Many of its founding fathers were white men who would talk about all that they love in wildness, but then, in the next breath, dig into Indigenous gravesites, fondle the skulls of people and claim that they were less intelligent because of their skull shape, as well as promote the enslavement of people. Some folks want to excuse it all away and say these were men of their time. John

James Audubon and his contemporary Henry David Thoreau were two white men of their time but with very different views. In Audubon's story entitled "The Runaway," he ran across a Black man in the woods who'd successfully reunited his family, but instead of helping them find freedom, Audubon decided to turn the family back over to their master, claiming they were happy to be re-enslaved. Thoreau, by contrast, was a man who was helping self-liberated Black people escape to freedom at the same time *Walden* was being published. During that period, he railed against the Fugitive Slave Act in Massachusetts because a Black man had been taken off the streets and re-enslaved. Thoreau was pissed off about that. Today, it amazes me how often people, especially white men, talk about freedom and liberty, but then they want to deny it for me. You want to run a rattlesnake flag up a pole and say, "Don't tread on me," even as you got your foot or knee on the neck of somebody who does not look like you. All that comes together for me in conversations about conservation. I can't separate race issues from conservation issues. To separate race and conservation makes them small, and some would say to make them small makes them palatable. Well, guess what, I don't

To separate race and conservation makes them small, and some would say to make them small makes them palatable. Well, guess what, I don't want all of this to be palatable.

A VIEW OF THE FOREST AND LAKE DURING TWILIGHT AT DREW'S SUNSET CAMP
BY DUDLEY EDMONDSON

want all of this to be palatable. I don't need you to swallow and forget the pain and misery of other people. I need that to stick in your craw. And I need you to cough on it, and I need you to figure out how to make it better.

ENVIRONMENTAL JUSTICE IS A CIVIL RIGHT

People want to talk about environmental justice as something separate from civil rights. I don't separate the two. Clean air, clean water, and clean soil in which to grow healthy foods should not be a privilege; they're a right. Conservation communities and the organizations that are tasked with "saving nature" have so separated people from nature that there's a big blind spot there. That blind spot leaves out many people who have to be included if conservation is to be a far-reaching effort. Otherwise, it's going to continue to only be privileged people talking about wild places that they can use their privilege to go and visit.

Climate change is a precipitous edge that's not only going to impact wild beings, but is also going to impact humanity. If you are a Black person, Brown person, or Red person in this country, or impoverished white person for that matter, you're often going to suffer in ways that others don't see, or others can't seem to feel. There's an old saying that goes, "When America gets a cold, Black folks get the flu, and if, America gets the flu, Black folks get pneumonia and can't breathe at all." I think it's that way with climate change, in that these groups are first on the chopping block because it's going to impact where we live first. Issues of climate change exacerbate everything else. It's like an exponent on misery.

DREW SPENT A CONSIDERABLE AMOUNT OF TIME LIVING WITH HIS GRAND-MOTHER, MAMATHA, AT HER COUNTRY HOME.
COURTESY OF DREW LANHAM

DREW STANDS ON THE UPPER DECK OF HIS SUNSET CAMP CABIN.
BY DUDLEY EDMONDSON

LEARN MORE

Joy Is the Justice We Give Ourselves, Hub City Press, 2024

Sparrow Envy: Field Guide to Birds and Lesser Beasts,
Hub City Press, 2021

The Home Place: Memoirs of a Colored Man's Love Affair with Nature,
Milkweed Editions, 2017

Like other folks in this book, Drew and I have known each other for many years. We even have the same birthday. Drew is a great orator and a celebrated author. His unique way of looking at the natural world can make some people in the conservation community uncomfortable, as he points out our nation's continued struggle with historical race issues and its ideology that people and nature should remain apart. Like so many in this book, Drew's viewpoints are a departure from conventional thinking, but they are how we will need to think if we are to have a promising, inclusive future that allows us to care for others and repair our relationship with nature.

—DUDLEY EDMONDSON

CHAD BROWN

FOUNDER AND PRESIDENT OF SOUL RIVER INC.

PORTLAND, OR

Chad Brown and I are friends and filmmakers with a focus on nature and the environment and turning the camera more on folks of color. We're also both adventurers. We like remote places, and we traveled together with a group to the Arctic National Wildlife Refuge, where we traversed across tundra and learned about wildlife and the Indigenous residents.

—DUDLEY EDMONDSON

CHAD BROWN WITH HIS SERVICE DOG, AX, IN THE ARCTIC NATIONAL WILDLIFE REFUGE
DURING ONE OF HIS LOVE IS KING EXPEDITIONS
BY DUDLEY EDMONDSON

AN ARTIST AND A CONSERVATIONIST

I'm the founder of two nonprofits, Love is King and Soul River Inc., and I am also a photographer, a conservationist, a filmmaker, and a graphic designer. I wear many hats, really. My nonprofit Love is King aims to eliminate fear and establish safety for Black, Indigenous, and all people of color while on our public lands. It is committed to creating a reimagined outdoor environment focused on participation, representation, inspiration, and advocacy.

The mission of Soul River Inc. is merging at-risk youth of color with combat and disabled vets and bringing these two worlds together so they serve one another. It gives the veterans purpose as they help our youth build leadership skills and confidence. Our ultimate goal is raising youth to be environmental leaders so they can learn how to protect our public lands and also take a stand in supporting Indigenous communities.

AS A CHILD, BLACK COWBOYS, FARMERS, AND THE OUTDOORS WAS ALL I KNEW

I come from a family of farmers and cowboys and ranchers. My father was a hunter, and he worked the land with my grandfather and my grandma. We ran the Brown Ranch, a cattle ranch in Cuero, Texas. I used to spend a lot of my summers as a little kid growing up around my dad, grandpa, and grandma. I'd wake up in the morning to a rooster crowing. I'd sit at the coffee table and drink coffee with my grandpa. I was too young to drink out of coffee cups, so he'd pour coffee in a saucer, and I would sip it. I'd walk out in the morning with my grandpa, and we'd feed the hogs, and he'd get the eggs from the chickens, and those eggs would become breakfast. Just experiencing the whole summers with my grandparents, I got great memories. I remember

CHAD (RIGHT) WITH HIS BROTHER, NEXT TO THE BROWN ROAD SIGN ON HIS FAMILY'S LAND
COURTESY OF CHAD BROWN

CHAD'S PET DEER, BAMBI, IN THE ONLY PHOTO
HE HAS OF THE ANIMAL
COURTESY OF CHAD BROWN

acting like I was Daniel Boone. My grandpa made a little raccoon hat for me. I would have my little .22 rifle with me, and we would walk out together and hunt.

One day, when I was not with him, he came across a dead deer with a fawn beside it. He picked up the fawn, brought it home, and that became my pet. When I went back to the city after summer break, I was known as that kid that walked his deer around the neighborhood on a dog leash in San Marcos, Texas. I raised Bambi, and they allowed me to bring it into the classroom where it would curl up next to my desk. I was a kid with a deer, and it was really, really cool.

I was exposed to a world of Black folks that were ranchers and cowboys, that's all I knew. I have great memories of my dad and his brothers entertaining us on a Sunday. They would be out there wrestling bulls in front of us and bringing cattle in. I was exposed to Black rodeos. That was my upbringing as a child, very rich with outdoor experiences.

MY STRUGGLES WITH PTSD HELPED ME REALIZE THAT THE OUTDOORS IS A HEALING PLACE

When I went off to college, I attended art school. That took me away from the outdoor environment for a long time. I couldn't

finish art school because of money, so I went into the military to get the GI Bill, to complete my schooling. I joined the service and was part of a joint Task Force Expeditionary Unit in the United States Navy. I served in Desert Storm/Shield in the Gulf War and Operation Restore Hope in Somalia. That took me all over the world, through 14 different countries. My time in the service really crippled my mind to the point where I ended up becoming 50% mentally disabled with PTSD. I also have a 20% TBI, which is traumatic brain injury, from being around too many explosives.

There definitely are many scars from my time in the service, but I still used my GI Bill to finish my schooling. I got a bachelor of fine arts in commercial art and design. I took that to New York to follow my dreams, just like a lot of creatives.

CHAD GRADUATING FROM PRATT UNIVERSITY IN NEW YORK, PICTURED WITH HIS MOTHER (LEFT) AS WELL FACULTY AND FELLOW GRADUATES
COURTESY OF CHAD BROWN

While I was there, I got my master of fine arts at Pratt Institute in visual communication with a focus on photography, art direction, and advertising. That launched me into working for ad and design agencies across the United States from L.A. to New York. Later, I got a senior art director offer from an ad agency in Portland, Oregon.

Coming to Portland was like moving to a country town in comparison to the pace of New York, so I had a lot of extra time on my hands. As a result, my mind started to relapse, and I began thinking about stuff I had seen and been through in the wars, and I was forced to deal with it. That

crippled me, and over the course of about 6 years, I had to fight my demons, and I attempted suicide a couple times, and that put me back in the VA hospital. I had to put my whole creative career on hold and deal with my mental health.

While I was in this very dark, dark mental space, a new friend came into my life and took me to the river and put a fly rod in my hand. That is how I was able to cope with the stuff I was going through. I got my first fly rod and came back to the Clackamas River in Oregon. I didn't know what I was doing, but one day, I hooked into a jack (male) salmon and started hooting and hollering. I will never forget that day, because I was so strung out on many medications to help me fight depression and keep me from having bad dreams. That salmon really got me stoked and excited. It felt like new medicine. In fact, my VA doctor strongly encouraged me to fish more. As I continued to fish, my community changed, and I found myself spending more time with anglers, hunters, and conservationists. I was really reestablishing myself with nature, realizing the outdoors has a deeper meaning, which is that it is a healing place.

THE OUTDOORS CAN BE A PLACE TO NURTURE RAW LEADERSHIP IN YOUTH

When I stepped back into society, I wanted to channel a lot of the things that I'd learned on the river and through this community into my first organization, Soul River Inc. I wanted youth and veterans who were troubled and are displaced in their own communities, fighting depression, to know that the river is not just there to go and swim and recreate, but it's also a place of healing. That became the organization's mission. It also became the building blocks of my professional career in conservation and environmental justice.

As I mentioned earlier, the mission for Soul River Inc. is to connect inner-city BIPOC youth and US military veterans to the outdoors through many different outdoor educational and transformational experiences. We do what we call deployments, and they range from identifying environmental issues happening on public lands, around wildlife, in freshwater, and the concerns of Indigenous peoples.

DEPLOYMENTS INSIDE THE ARCTIC CIRCLE

Some of our biggest deployments are within the Arctic Circle. Youth and veterans get a chance to immerse themselves in Indigenous culture within the Arctic wilderness. They get a chance to connect with the land, hike the tundra, and see wildlife. Through that process, they face some pretty big challenges. On one deployment in particular, we were traveling the Ivishak River, which flows south from the north side of the Brooks Range and onto the Arctic tundra. The Ivishak is a difficult river to navigate because it is veiny, with many different channels. That day, we were doing a 260-mile run in expedition canoes. Each boat had two youth, one in the stern steering and the other in the bow position, navigating the boat. The veteran is in the middle and kind of steps back and lets the youth take the lead and learn how to navigate wilderness water. If, for some reason, the youth calls the wrong shot, the vet is there to help get them back on course. On this particular day, the youth navigator made an error at a split in the river, and their boat went down one channel, and the rest of us took the correct channel. Very quickly, they faded out of sight, and we lost radio communication with them. The weather was nasty and rainy with 25-30 mph winds coming upstream. We pulled over on the bank where there was an inlet back to the main channel, and we waited. Several hours passed as we waited, wet and cold in the rain, hoping they were safe. I was about to get on the sat phone and call for an emergency rescue when, out in the distance, we heard a strong stroke cadence being called by the veteran from the lost boat. The youth had to dig deep and fight the current and high winds. We couldn't see them, but we heard the vet's voice. It was only the three of them working together as a team, fighting against nature. We finally saw them in the distance, fighting the elements. We started rooting and cheering from the riverbank.

To me, I felt like I was witnessing leadership in its rawest form, watching these youth working together with the vet, seeing this situation through. They were in a very serious situation, paddling upstream in high winds, but they made their way across that water to us as we all celebrated.

> To me, I felt like I was witnessing leadership in its rawest form, watching these youth working together with the vet, seeing this situation through.

THE FUTURE OF CONSERVATION IS IN THE HANDS OF YOUTH OF COLOR

That's the nuts and bolts of the organization: nurturing that raw leadership that we believe in inside every young person. By inserting them into situations that give them opportunity for growth, we hope they'll be able to step up and become environmental ambassadors. The hope is it will build confidence as they reflect on their wilderness experiences, knowing if they can pull through that, they can pull through anything. When a lot of these young people come back, they do continue building their leadership skills. One youth from a past deployment is studying environmental law at Oregon State University.

When we come back from a deployment, we often take it to the next level by setting up meetings with members of congress. This gives our youth an opportunity to learn how to navigate hard conversations when it comes to protecting the environment and the rights

ONE OF THE VERY FIRST SOUL RIVER DEPLOYMENT GROUPS, IN ALASKA ON THE IVISHAK RIVER
IN THE ARCTIC NATIONAL WILDLIFE REFUGE
COURTESY OF CHAD BROWN

of Indigenous communities. They get an opportunity to come in and express themselves and share their thoughts and experiences. The underlying strategy is persuading congressional members to take a second look by considering the wisdom of these young people. Several vets attend these meetings, but we basically act as chaperones and give support and encouragement by letting them know we have their backs as they grow their leadership skills. That's the community part of the organization, how we step in and how veterans work with youth. These soldiers fought for this country and came back, and while the troubled youth are not soldiers, they're both fighting for their lives, and they need that outdoor space and the support of a community to help them find healing and purpose again.

LEARN MORE

Soul River Inc.
soulriverinc.org

Love Is King
loveisking.org

Chad and I have known each other for a few years now and have traveled extensively throughout the Arctic. We understand the healing power of remote wilderness spaces. Chad wears many hats—he's a nonprofit director, a photographer, and filmmaker—but his dedication to youth and veterans, by providing them opportunities to connect with nature for better mental health, is always at the forefront of his mind. —DUDLEY EDMONDSON

AREAS WITH NON-WHITE POPULATIONS ARE OFTEN SUBJECTED TO POLLUTION FROM HEAVY INDUSTRY
BY SSISABAL/SHUTTERSTOCK

WHAT?!

"Areas with low-income groups have been consistently exposed to higher average levels of particulate matter air pollution."

—JBAILY, ET AL., *NATURE*, 2023

RUE MAPP

CEO OF OUTDOOR AFRO

OAKLAND, CA

Rue Mapp runs Outdoor Afro and has been doing amazing things. We both believe in the importance of connecting Black folks to the outdoors and reconnecting Black folks and folks of color to nature and getting people to see folks who look like them as advocates for nature and the environment.

—DUDLEY EDMONDSON

RUE STANDING AMONG THE CALIFORNIA REDWOODS
BY ROBIN BRUMFIELD-JOHNSON

BLACK CONNECTIONS AND LEADERSHIP IN NATURE

I started Outdoor Afro to inspire Black connections and leadership in nature. Today, we are a national not-for-profit organization with leadership networks around the country in 56 cities, connecting thousands of people to outdoor experiences. My work has been focused on Black people because there's a unique Black narrative in this country that we have to find our way toward healing through. Our bodies swung from trees, so the idea that Black people are going to be tree huggers flies in the face of what the historic Black experience has been for some and what echoes in our consciousness.

RUE AS A CHILD IN CLEARLAKE, CALIFORNIA
COURTESY OF RUE MAPP

If we don't repair the severed relationship between Black people and the natural world, we're not going to care about nature. I recognized early that Outdoor Afro's work had to be highly relational, meaning we're not going to rush anyone to the altar of conservation and tell people to care about something that they've never experienced in a pleasurable way. For example, when I see children outside cleaning up a trail or picking up cigarette butts off of a beach, my first question is, has this child ever played there before? Have they become familiarized with this place before? If their first introduction to nature is work, that is unfair, and it is a way our community has been misused in traditional environmental outreach. If a child does not know how to swim, then they are not going to do a whole range of outdoor activities. They're not going to put a pole in a lazy lake. They're not going to ease into a tippy kayak. And they're absolutely not going to give a damn about plastic in the ocean. So how do we get people to care about something that they don't have a relationship with? I think that's the role that Outdoor Afro gets to play in helping people develop a trusting, loving, and inspired relationship with the outdoors.

MY CHILDHOOD ON OUR FAMILY FARM

I had a unique upbringing, but I didn't know it was unique until I became an adult. My parents, who were from the Deep South, cared so much about their relationship with the outdoors that they set up a family ranch space about 100 miles north of Oakland. At the ranch, we would host celebrations and do all kinds of activities that were about having our hands on the land. I would work in the garden and pick fruit. We had domestic animals on the property we used for meat, but we also hunted, and it was a family affair to bring those foods to the table and to sustain the family. I didn't realize how special it was until I got much older and had a yearning inside of me to tell a new narrative about Black people in the outdoors and lift up that aspect of my family heritage. As I talked to more people who look like me, I recognized that while I had a unique experience while living in the Bay Area, there are parts of my story in almost every Black family in this country that mirror that same deep connection: a nature know-how, that today I lift up more in my own life, but also share those stories and experiences with others.

ENVIRONMENTAL JUSTICE THROUGH HELPING PEOPLE UNDERSTAND POLICY

My work is focused on Black America because I want to speak to the work of our ancestors and do everything I can to lift up the

RUE WITH A GROUP OF OUTDOOR AFRO PARTICIPANTS AT AN EVENT
COURTESY OF RUE MAPP

everyday person in your family. Those who taught you how to garden, whoever taught you how to fish, and taught you how to love and put your hands on the land—those to me are the nature teachers that we need now more than ever. Another thing Outdoor Afro has done is that we've put together a civics learning community, and that learning community has actually gone to Congress and the Senate to meet with our local elected representatives, and have made it clear over the last few years that Outdoor Afro has a community of people who care about laws that protect the environment. I think it's been important for us to elevate our voice in those places. But the other piece of this is ongoing education. My policy director

believes that it is her goal to make sure that people are able to hear information about policy with language that my Auntie would understand. That's why I also serve as a California State Park commissioner, because I think that there's a big disconnect between a lot of the policies that are happening and how it matters to communities. These are the ways we show up, even though we do not deliberately refer to ourselves as an environmental justice organization.

OUR WORK IS ABOUT HELPING PEOPLE HEAL

There has been a lot of civil unrest around the topic of police involved violence, and I asked myself, what am I supposed to be doing right now as a leader of a Black-focused organization telling people to go outside? What am I supposed to be doing and talking about right now? The answer just came to me. It was so clear. It was, "You do nature Rue, that's your lane." That weekend I pulled together a group of friends and partners and various folks from the Outdoor Afro community. We went into a meadow among the redwoods, and we began with our breathing, and stretched our bodies under the morning sun. When we got on the trail, and headed down into the forest, I became so much more aware that the conversation around me felt lighter and anything that anyone was carrying in the way of stress was absorbed into that forest. That was a big moment for me because I realized our historic connections—we were doing as the song goes, laying down our burdens, "Down by the Riverside" as we got down in that redwood bowl along the stream trail. Black people have always known where we could take our burdens and we were following that tradition. It was in that moment that I knew my work was about healing. I always say, at Outdoor Afro we celebrate and inspire Black leadership and connections in nature, and we do this because we want people to take better care of themselves, take better care of their communities, and ultimately help us all take better care of our planet.

> I always say, at Outdoor Afro we celebrate and inspire Black leadership and connections in nature, and we do this because we want people to take better care of themselves, take better care of their communities, and ultimately help us all take better care of our planet.

LEARN MORE

Outdoor Afro
outdoorafro.org

Outdoor Afro Inc
outdoorafro.inc

Nature Swagger, Chronicle Books, 2023
chroniclebooks.com/products/nature-swagger

I think Rue and I began our mission to get more Black folks into the outdoors around the same time. Her work started in Oakland, California, and now there are Outdoor Afro groups all across the country. Rue is unapologetically Black, and that is what I think has made Outdoor Afro such a success in larger cities across the country. —DUDLEY EDMONDSON

CHRISTOPHER KILGOUR

FOUNDER: COLOR IN THE OUTDOORS AND COMMUNITY OUTREACH MANAGER FOR THE NELSON INSTITUTE FOR ENVIRONMENTAL STUDIES

MADISON, WI

Christopher Kilgour and I do a lot of work that is very similar— working with folks of color to help understand nature and the environment, learn outdoor skills, and enjoy recreational activities. Like me, he understands how healing having a connection to nature is, and he recognizes that there's a sense of safety that comes with being on a person of color's land that you don't feel on public land.

—DUDLEY EDMONDSON

CHRISTOPHER ON THE EDGE OF HIS 92-ACRE PROPERTY NORTH OF MADISON, WISCONSIN
BY DUDLEY EDMONDSON

COLOR IN THE OUTDOORS

I'm the founder and director of Color in the Outdoors. It's an organization that was born to provide access, safe space, and engagement in outdoor spaces. To coin a phrase from a woman named Hazel Symonette, who is a professor at the University of Wisconsin-Madison and one of the wisest human beings I've ever known: "Existence is a form of resistance." I found that to always resonate and realized that I was doing just that. Not only am I stubborn enough to be in "these spaces," but I feel that I'm capable and competent enough

CHRISTOPHER WITH HIS SISTER, SARAH, ON HIS SHOULDERS IN JAPAN
COURTESY OF CHRISTOPHER KILGOUR

to hold my ground—not necessarily in a confrontational way—but definitely in a defiant way. That's one of our founding mantras: to take up space unapologetically.

MY PARENTS EXPOSED VAN-LOADS OF KIDS TO NATURE

I grew up near Tenney Park in the center of Madison, Wisconsin. It's a gem in the midst of a city, and it felt like a little sanctuary. It has transformed over the years, but that was my playground as a child. And my parents were a big influence. They would say, "Get out the house; go outside!" They instilled the love, respect, and understanding of stewardship in outdoor spaces. My parents were willing to take vanloads of kids everywhere we went. Partially because it would keep me busy, but also, I think they understood the value to that, and so I was able to have those adventures. I was also very fortunate to have other adults and peers who spent time in outdoor spaces and were willing to share their love and knowledge of those spaces, which just kind of continued to grow mine.

At a young age, I did see that it was predominantly my white friends that were out there doing those things. My Brown friends were often saying to me the same things that I continue to hear to this day,

CHRISTOPHER WITH FORMER MADISON POLICE CHIEF RICHARD WILLIAMS
COURTESY OF CHRISTOPHER KILGOUR

which is, you know, "Black folks don't do this. Latino folks don't do that, etc." That's kind of what was really my driving force. Part of it was just selfish, that I wanted to have more of my friends around, and I wanted to do more things with them. And the only way to do that was to get them out with me in outdoor spaces. Also, on kind of generic levels at a young age, we'd have these conversations about, "Isn't it a shame that there's not more of us out here? Isn't it messed up that so and so, our friend from middle school or high school, either, doesn't want to? Or, they want to, but their parents won't let them because of warranted fears from previous bad experiences or just from being socialized that outdoor space isn't for us? Also, they were being told that if you go out to these spaces, you not only may encounter dangerous animals or a dangerous environment, but you, more importantly and more terrifyingly, you will encounter dangerous humans.

MY RACIAL IDENTITY COMES WITH AN ASTERISK

My identity, in general, as someone who was adopted, was something that I struggled with as a young kid because the Brown folks said that I wasn't Brown enough. The white folks said that I wasn't white enough. The Mixed folks said, well, you aren't truly as Mixed as you could be because you don't necessarily know your heritage 100%. I talk about my connection to the native community, but I can't, for instance, definitively say I belong

THE COLOR IN THE OUTDOORS EDUCATIONAL BUILDING ON CHRISTOPHER'S 92-ACRE PROPERTY NORTH OF MADISON, WISCONSIN
BY DUDLEY EDMONDSON

to this governmentally recognized or designated tribal entity. I can say that I abide by and believe in a lot of the ideologies that are considered part of Indigenous teachings, community, and culture. I also very much celebrate, acknowledge, and hold dear my African American roots and my potential Pacific Islander roots. That is something that plays a strong role in how I walk through the world and, ultimately, being very cognizant of just being one's own person.

Another challenging thing that I have dealt with was just that I was always a big dude. When I was in high school, I was 6'3", 195 pounds, and so everyone always expected

that I was going to be "the protector." I did become a bodyguard and worked in law enforcement as an adult, but that also relates back to why nature has always been so important to me. It was one of the few spaces where I could unload, where I could let that stress out from work, where I could be me and not be ashamed or afraid to just be.

COLOR IN THE OUTDOORS, BY US FOR US

The birth of Color in the Outdoors happened when I was still a teenager. I just couldn't conceptualize how to make it into a more organized business entity. I was thinking that there needs to be a space and a vehicle, literally and figuratively, to get folks outdoors, but not just "let's go on a field trip" kind of mindset, but how can we do this in a sustainable and meaningful way and have a long-lasting impact? At times, Color in the Outdoors was like the casual pick-up game: Tell your friends we're going on a hike and to be there. Pretty soon, it became more of a group versus just kind of going to do a tour somewhere with somebody else.

As the field trips continued, I asked the participants, "In a perfect world, would you see this being a thing? How might it be different than what you've done with other groups in other situations?" That's where I really started to see that the outdoor activity, recreation, and engagement space can be toxically competitive at times, and unless you really had your shit together, you were constantly playing this kind of Oliver Twist-esque role of walking in with your hat in hand and begging somebody else for money, resources, or access. It was never that you're actually part of the conversation dynamic. It was more, "We will allow you, and we're going to make sure you understand we're allowing you to occupy this space for this short period of time." This has been going on since

the occupation of this continent by folks other than Brown people.

HUMANS HAVE CONSISTENTLY TRIED TO CONTROL NATURE, AND IT CONSISTENTLY REMINDS US THAT'S NOT HAPPENING

That's where I really saw this struggle with trying to code switch and having to constantly change your tone, your demeanor, your vernacular, depending on who is in the room. That in and of itself is exhausting. But the performative nature of that was just too toxic; I just couldn't do it.

Now, we have a 92-acre retired iron mine being reclaimed by nature as the home of Color in the Outdoors. Humans have consistently tried to control nature, and it consistently reminds them that's not happening. The property is forested with prairie and agricultural space, with a lake in the middle. We want to be able to provide agricultural space for people that don't have access to a growing space. These habitats could then also be utilized as research projects for students. The goal is being able to provide educational opportunities. We also teach outdoor skills like camping, fishing, hunting, and more. Instead of constantly trying to figure out how to do it or get it based on somebody

A GROUP OF STUDENTS AT A COLOR IN THE OUTDOORS HUNTING TRAINING CLASS
COURTESY OF CHRISTOPHER KILGOUR

else's definition, I've decided we need to do this by us, for us.

HAVING A STROKE STRENGTHENED MY CONNECTION TO NATURE

Strength, both physical and emotional, really, and rather unfortunately, took its toll on me, especially working in the law enforcement world. I was 26 at the time, and we had to be stronger, faster, and have longer endurance because it literally was a matter of life and death in some cases. We're in the gym all the time after we would do these search warrants or whatever. That night in particular, I had a headache when we were getting done with work, and I didn't think

much of it. I thought I was just dehydrated or something instead of thinking something was maybe wrong. We were pushing lots of weight close to a 500-pound bench press for reps. I had a good workout and went to get some breakfast at a local joint before going home. The headache kept getting worse.

The next morning, I woke up with a thundering headache. My girlfriend at the time asked me what's wrong. I said, "I don't know." I went downstairs, got some orange juice, and went back upstairs to tell her it was time to get ready for work, and I passed out cold on the bed. When I finally came to, she was pushing me to get up, saying, "Stop playing." I said, "I can't move; I can't open my eyes." She said, "Your eyes are wide open." I was like, "Well then, we've got a bigger problem." At that point, she freaked out, and she knew something was wrong when I was trying to talk, and one half of my face wasn't moving. She was able to drag me down the stairs and kind of get me to the truck and take me a few blocks down the street to our clinic. They came in to take a look at me and they said, "You really need to get to the hospital as quickly as you can, don't wait for an ambulance." By the time I made it to the hospital, I had crashed. I was in pretty bad shape. I couldn't really do much. Years before that, I had gone through EMT training. One of the things that our instructor kept saying over and over is never anticipate that the person you're working on can't hear you when you're doing CPR and they're not responding. Don't ever address them or talk about them as if they're not there.

I had a moment like that during my stroke. I distinctly remember hearing the doctors and the nurses talking and hearing the noise of the heart monitor—beep, beep, beep. I could even hear the long beeps from the heart monitor, then it started again as they were pounding on me, trying to get me to respond.

They ran a barrage of tests and never figured out what had happened. They said that there may have been some blockage but that they couldn't find it. But there was clearly some damage to my brain, and that the likelihood of me ever walking again or having unaffected vision was minimal at best. They said, "Because we don't know what happened, this could happen again, and if it does, it will probably kill you." My response to that statement was pretty direct, and maybe I used some explicit words, but I told them, "I am not done here yet." Later, the doctor said that that attitude about living is probably one of the things that brought me back.

Later, the doctor said that that attitude about living is probably one of the things that brought me back.

I MET A CARDIOLOGIST WHO LOOKED KIND OF LIKE SANTA CLAUS. HE GAVE ME MY LIFE BACK.

Over the course of the next almost nine months, I couldn't remember my parent's names. I knew them, but I couldn't quite figure it out. Eventually, my memory came back. My vision slowly started to return. Actually, to this day, if I look at a pie chart, the upper right quadrant of that pie chart doesn't exist anymore.

Fast-forward to 2000. I was helping some friends, doing some construction work on the side. I was on my way back to my house when I had another stroke. This time it was not as bad, but I was in rush-hour traffic. It felt like someone hit me in the back of the head with a cattle prod. I was able to kind of direct the truck back. Luckily, I wasn't too far from home, so I kind of just rolled the truck up on the lawn and honked the horn because I knew my cousin was home, and he came out. I said, "Dude, I need you to take me to the hospital." Luckily, that time they did more testing, and they found that I had a hole in my heart that had always been there. It's a condition called patent foramen ovale (PFO), a hole between the left and right atria (upper chambers) of the heart. They say about 30% of the population has these, and many folks live their entire lives without them ever actually being an issue. What wound up happening is under heavy exertion, it can actually cause that valve to open and cause blood clots that would normally go to your lungs and dissolve to go to the other side and get up into your brain, which is what a stroke is.

> I was on my way back to my house when I had another stroke. It felt like someone hit me in the back of the head with a cattle prod.

I met with a cardiologist, who looked kind of like Santa Claus, here in Madison. He was one of the tops in his field. He slid a pad of paper across the table and said, "I want you to write down all the things that you like to do." Of course, 90% of it was outdoor activities or outdoor-related. Then he said, "Well, here's how this works: we're going to have to patch this hole. We split your chest open, open up your rib cage, and put in a Gore-Tex mesh patch." He said, "That seam in your chest and that bone will fuse together, but you will always run the risk of opening it up again under heavy exertion. Also, you'll never be as strong as you once were. There is another option. There's a couple of physicians who work over in Milwaukee, Wisconsin, and they're just revolutionizing a new process in the Pediatric Medicine unit right now, where they go through your femoral artery in your leg and do these repairs with an arthroscopic procedure. They never have to crack the chest."

I didn't get in for surgery until late winter, early spring of 2003. The reason I know that date as well as I do is that I found out that my girlfriend at the time, who was about to be my second wife, was pregnant with my son. I wanted to get this done before he was born.

CHRISTOPHER TAKES A SELFIE WITH ROXIE HENTZ (LEFT) AND DONNETTA FOX (RIGHT) AT A COLOR IN THE OUTDOORS EVENT
COURTESY OF CHRISTOPHER KILGOUR

I wanted to be there for him at 100%. That surgery fixed it. Having two strokes, then heart surgery on the eve of the birth of my first child—those were moments where you ask, "What is a priority to me?" They're epiphany moments where you say, "Am I doing things the way I should be? Where am I putting my time, my energy, my focus? That really shifted things and saved my trajectory on some levels and allowed me the opportunity to continue to do the things I love to do.

I FEEL LIKE I JUST NEED TO SIT STILL AND BREATHE AND SMELL NATURE

When you finally slow down enough, you realize how important it is to slow down and understand what that actually means. My relationship with nature just drove that home even more. Now when I go out to the woods, it's legitimate, and I feel like I just need to sit still and breathe and meditate and literally hear, see, and smell nature.

My friends to this day tease me all the time; they say, "You have superhuman senses."

Decades ago, I shot a deer with my bow, and it ran into some thick brush. I called one of my buddies, and I said, "Hey, if you're around, help me track this deer before I lose daylight and lose the deer." But before he even got there, I called back and I said, "Hey, I found it." He said, "You saw it?" I said, "No, I can smell it." He's like, "No way, what?" I said, "I can smell it; the wind shifted. I can smell blood." He's like, "Dude, you can't." I said, "Okay, stay on the phone with me." I walked in on the blood trail; I literally caught that smell in the air.

My stroke, it changed me. I didn't get bit by a radioactive spider, but being on that gurney, on that emergency room table, in that space, I understood what it was to be still. That really did heighten my senses. It impacted how I walked, literally and spiritually, through the world, especially in outdoor spaces.

LEARN MORE

Color in the Outdoors
colorintheoutdoors.com

Christopher dreams big and delivers big. When he created his outdoor education facility, located on a large tract of his land, he deliberately did so to provide better access to environmental education opportunities in a space that centers people of color first. So he built a facility to teach BIPOC how to hunt, fish, camp, and more. His mission is help them better understand the natural world. —DUDLEY EDMONDSON

A PRIVATE LAKE ON CHRISTOPHER'S PROPERTY WHERE SOME OF THE COLOR IN THE OUTDOORS EDUCATIONAL PROGRAMMING OCCURS
COURTESY OF CHRISTOPHER KILGOUR

NICOLE JACKSON

ENVIRONMENTAL EDUCATOR AND FOUNDER OF N HER NATURE LLC

COLUMBUS, OH

Nicole Jackson and I share this passion of wanting to connect Black and Brown folks to the nature and the environment, especially because of the mental health benefits it offers. Her focus is ensuring that Black women can find opportunities to learn new skills, have high self-esteem and confidence wherever they are, particularly in outdoor spaces.

—DUDLEY EDMONDSON

NICOLE LAUGHS DURING A CONVERSATION AS SHE STANDS ON THE BOARDWALK
AT GRANGE INSURANCE AUDUBON CENTER IN COLUMBUS, OHIO
BY DUDLEY EDMONDSON

NATURE AS THERAPY

My professional background is in environmental education and interpretation from The Ohio State University. I'm also currently working as a nature coach for my business, N Her Nature LLC, which focuses on nature therapy for Black women. When I'm in nature, I feel centered. I feel at peace. I feel like I can learn a lot of things without having to verbally express it or say it out loud. And it's a very liberating space to be in.

GROWING UP IN CLEVELAND, MAKING A LOT OUT OF A LITTLE

Growing up in Cleveland, Ohio, I had a lot of different interests, but I used nature as an educational tool. My siblings and I lived in inner-city Cleveland, so there's not a lot of "nice" green space since the city is industrial and has many vacant lots. But even just seeing some patches of green was good enough for me to feel connected to nature.

I spent a lot of time out in nature by myself but did enjoy being outside with my siblings. We would go to local parks within walking distance, spend time in our backyard building things like forts, and would create outdoor plays. We learned to make a lot out of the few resources we did have.

ADVOCATING FOR BLACK WOMEN'S MENTAL HEALTH THROUGH NATURE

In 2021, I left my program coordinator position at the university. I decided to start my own business that focused on creating meaningful experiences in nature for Black women so they could better prioritize their mental health and well-being. I realized after COVID began that I wasn't taking care of my needs. I was running on empty and needed to pivot. I wanted to provide a way for Black women to create their own self-care toolbox, to empower them to be better versions of themselves.

NICOLE (CENTER) WITH SIBLINGS AND COUSINS
COURTESY OF NICOLE JACKSON

I felt like getting Black women out in nature provides a great opportunity to find solutions to mental health concerns, as well as creates more joyous and educational moments in outdoor spaces. This belief led me to provide birdwatching programs, nature walks, and other wellness activities. It's another way to help Black women see themselves in those spaces. They learn new things about the natural world, and it empowers them to prioritize their well-being.

The inspiration for N Her Nature came from reflecting on my mom, who raised 11 children alone and never had the chance to pursue her own interests. I had the chance growing up to explore and discover new opportunities and career paths. That's also something that I've always wanted for my mom. I know with Black women specifically, there's a lot of toxic stress in their personal lives as well as in their professional lives. If there were more ways for them to slow down and take better care of themselves, but also have fun, I feel they would have more balance and joy to celebrate. Since I didn't see very much of this growing up, I decided to set the standard for myself and be an example for others.

ENVIRONMENTAL EDUCATION, EVERYTHING IS CONNECTED

I've been doing environmental education for over a decade, and I started out at an internship at the Grange Insurance Audubon Center, in Columbus, Ohio. I worked as a summer camp counselor, teaching youth about the importance of urban nature and bird conservation. It was a fun opportunity for me to learn more about the city that I moved to. I was also teaching youth about the importance of accessing nature in their backyard. I feel like environmental education was something that I was drawn to because it's a career path that has a little bit of everything: math, science, wildlife, health, etc. I loved learning about the different connections between humans and nature, the benefits of being out in nature for your health, as well as topics like environmental justice, advocacy, and helping people understand that everything is connected.

I felt like getting Black women out in nature provides a great opportunity to find solutions to mental health concerns, as well as creates more joyous and educational moments in outdoor spaces.

THE IMPORTANCE OF STORYTELLING IN ENVIRONMENTAL EDUCATION

Many of the white-led organizations that I've worked for have

NICOLE LEADS A HIKE DURING ONE OF HER N HER NATURE EVENTS AT MOHICAN STATE PARK IN OHIO.
COURTESY OF NICOLE JACKSON

provided limited opportunities for training or professional development. Even the way I was teaching the importance of conservation and taking care of nature felt scripted. I wasn't connecting the storytelling to people's sense of place and how people connected to a green space geographically, historically, and even culturally.

Much of what was being left out was an explanation of the importance of having community involved in the care and stewardship of nature. That made me think about how I teach as an environmental educator and the importance of sharing my own story, how I got into birding, and how I fostered a loving relationship with nature. People assume that I've read all these books or started at a really young age with birding. And although my relationship with nature started when I was young, it started out as therapy for the trauma I experienced when I was in foster care around preschool age. It was a very dark time in my life. I was in survival mode. My love for learning came after that, and I got to build more joyous, educational memories for myself.

BEING AN INTROVERT IS LIKE HAVING A SUPERPOWER AS A NATURALIST

I think being an introvert is an awesome personality trait for a naturalist or someone who's interested in nature and the outdoors.

Nature is a space for you to really exist freely and engage your senses! There's a bunch of things happening around you that you can observe with awe and amazement. Experiencing nature is like a quiet noise. I know that's kind of an oxymoron, but that's the best way I can describe it. I'm an introvert. I used to think there was something wrong with me being more reserved and self-aware. I was always paying attention to the little things that other people would question and wonder, "How was that interesting to you?"

But now I see introversion as a superpower, especially doing work that entails connecting people to nature because it helps them see the extraordinary in the ordinary.

LEARN MORE
N Her Nature
instagram.com/nhernature

Nicole is a wonderful human being. Most of the time I spent with her during our interview in Columbus, Ohio, was spent laughing. We had so much fun talking about being introverts and our shared love of birds. Her work getting Black women into nature for better mental health is so crucial to the overall health of Black and Brown communities across the nation.

—DUDLEY EDMONDSON

ALEX TROUTMAN

WILDLIFE BIOLOGIST

AUSTELL, GA

Alex Troutman is a wildlife biologist, a birder, and an amazing human being. He talks a lot about how people have counted him out, and he's had every opportunity to prove people wrong and has done so. He's doing some amazing things in science and getting accolades for his work today.

—DUDLEY EDMONDSON

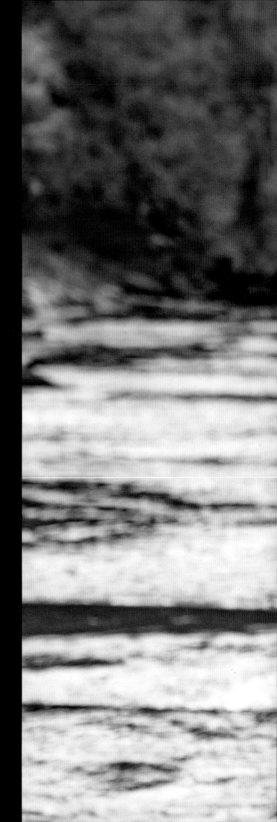

ALEX IN HIS U.S. FISH & WILDLIFE SERVICE UNIFORM, ON THE BANKS OF SWEETWATER CREEK
BY DUDLEY EDMONDSON

MY LIFE, MY WORK

I'm a wildlife biologist, working for the USFWS at the moment. Since I work with wildlife, habitat, and people, no two days are alike. In my role, I manage federal lands and monitor wildlife species to ensure population growth and stability. When managing or restoring lands, I get to use tractors to mow fields, trails, and waterfowl impoundments. We also do surveys before, during, and after large-scale land management activities to see how our actions have impacted wildlife.

ALEX AT 9 YEARS OLD; AS A CHILD, HE FOUND IT HARD TO CONCENTRATE ON YARD WORK WITHOUT BEING DISTRACTED BY ALL THE LIVING THINGS IN HIS YARD.
COURTESY OF ALEX TROUTMAN

A KID CONSTANTLY DISTRACTED BY NATURE

I grew up in Austell, Georgia. I was always interested in nature. Growing up, I was lucky to have had a creek in my backyard. We would spend a lot of time back there looking for water lizards, which we didn't know at the time were salamanders. We would also build up a dam to pool up the water so we could fish. Many times, my chores, like raking leaves or cutting the grass, were often extended because I would see a snake or a frog run in front of the lawnmower. I would chase and study it to see what it was. Or I'd spend time sifting through raked leaves to see what different grubs and things were hiding in them.

I spent many Saturdays fishing with my father, uncle, and brothers on a lake in rural Georgia. I saw my spark bird, the red-tailed hawk, while fishing there. The red glow of the tail soaring over and seeing the sun rays come through was memorable.

GROWING UP WITH NO ROLE MODELS THAT LOOKED LIKE ME

I grew up in the era of Jack Hanna, Steve Irwin, and Jeff Corwin, so for a long time, those were my idea of a wildlife biologist. Obviously, they're not Black, so I didn't have that true

SWEETWATER CREEK NEAR AUSTELL, GEORGIA
BY DUDLEY EDMONDSON

connection that this was something I could do. Originally, I thought I was going to be a veterinarian because that was one of the only representations of a Black person working with animals that I saw.

Growing up, going to zoos and national parks, I didn't see any Black people working as zookeepers or biologists. It wasn't until I worked for the National Park Service that I saw a person of color who was a biologist and who wasn't me. I was going to make it work regardless. I knew that there weren't any

Black people that I could connect with, at least so I thought. Thankfully, I found things like Black Birders Week and BlackAF STEM on social media platforms like Twitter and Instagram. People there had similar backgrounds to mine. It helped me realize there are many of us out there.

Through social media, I am able to show that Black people are scientists and enjoy the outdoors. I started showing what I regularly do as a Black biologist, whether it's just bird banding, working

A RED-TAILED HAWK IN FLIGHT, ONE OF ALEX'S FAVORITE BIRDS
BY DUDLEY EDMONDSON

with sea turtles, or looking for butterfly eggs. I hope that other people will see my work, especially young people.

THE TAINTED TABLE

A lot of people say there are plenty of seats at the table doing conservation and science work. But the people who built that table promoted and committed genocide and slavery through colonialist systems that treated people horribly. So even if we, as people of color, come and sit at that table, it is already tainted. If we come and do something good, the white people will take all the credit for it because they built the table. So that table needs to be destroyed, torn down, and burned, and a new table needs to be built by everyone coming together in equality. Getting rid of old ways of thinking and embracing coming change and Indigenous perspectives is important because they've been here longer and know best how we should reestablish habitat for wildlife.

> I truly believe that nature is for everyone, and we should all be able to have the same privileges that our white counterparts have without worrying about having the police called on us.

NATURE IS FOR EVERYONE

I truly believe that nature is for everyone, and we should all be able to have the same privileges that our white counterparts have without worrying about having the police called on us. Or even sometimes having to dress a certain way in order to not look too intimidating to white people. I've had many experiences where people have questioned my presence. One time,

I was working for the federal government in my full Fish and Wildlife Service uniform, and someone came up to me while I was nailing down a sign on federal property and asked me what I was doing, even questioning if I belonged in the area. I had to show them my federal identification to get them to leave me alone, but I should not have had to do that.

EDUCATIONAL MENTORS AND FRIENDS

I received a degree in biology from Georgia Southern University. It was during that time that I started taking field biology, ornithology, and mammalogy courses from two professors, Dr. Michelle Cawthorn and Dr. Ray Chandler. They became my unofficial mentors. They got me into the field, helping me to understand that studying birds was a career option. I ended up taking multiple classes with both. If I hadn't taken their classes, it would have been a lot longer before I figured out that wildlife conservation or studying wildlife was something that I could do. They helped me decide on being a biologist.

DON'T LET ANYONE TELL YOU YOU'RE NOT GOOD ENOUGH

One of the top things I'd like to do is help students like professors Cawthorn and Chandler helped me, even being a mentor to students who are struggling like I did. Someday, I'd like to have a foundation that bridges the gap for people of color and those of lower economic status to help them pursue careers through STEAM education.

During my education, I learned that professors have leeway in selecting students that they want in their lab.

AN EASTERN TOWHEE SINGING FROM A TREE BRANCH
BY DUDLEY EDMONDSON

Some would say to me, "Oh, looking at your GRE scores, I don't think you're the best fit for my lab," even though I had tons of experience. I know there are going to be instructors who will take the place of those who told me I wasn't good enough. Don't let them make you feel like you're not good enough. It's that professor who has the problem.

I've always been an underdog with the mentality of doing what people say I can't do. For me, it's about proving them wrong. I got my master's degree, so people couldn't say, "Oh, he scooted by in undergraduate school to get that degree." Being Black, while also having ADHD and dyslexia, made my two degrees even harder, but I did it.

I want to showcase to other people who may be in certain situations, or that come from lower economic status, who have learning disabilities that you can do anything you want, if you're passionate, go ahead and pursue it. Don't listen to the naysayers.

LEARN MORE

Critters of Georgia **(and many more),** AdventureKEEN, 2023

BlackAF in Stem
blackafinstem.com

> Nature has been a constant in Alex's life from childhood. He is an accomplished biologist and a pretty laid-back, but determined, young man. With multiple degrees in science, including a master's in biology, Alex is an excellent role model for future People of Color in science.
> —DUDLEY EDMONDSON

ALEX, IN HIS USFWS UNIFORM, HANDLING AN ALLIGATOR GAR
COURTESY OF ALEX TROUTMAN

THE ATIGUN RIVER NEAR ARCTIC NATIONAL WILDLIFE REFUGE
BY DUDLEY EDMONDSON

WHAT?!

"Historically, the United States has systematically segregated and excluded people of color from public lands and other natural places."

—"THE NATURE GAP," CENTER FOR AMERICAN PROGRESS, ROWLAND-SHEA, ET AL., 2020

SIQIÑIQ MAUPIN

ENVIRONMENTAL JUSTICE ACTIVIST

FAIRBANKS, AK

Siqiñiq Maupin is a huge advocate for Indigenous folks in the
Arctic and is not afraid to talk
bluntly about hard subjects like
cancer clusters near oil fields
in the Arctic and the serious
health and climate impacts from
resource extraction in the Arctic
Slope and around the world.
They have traveled nationally
and internationally to get the
story out about her people,
how industry and pollution is
destroying the tundra, and how
animals like caribou and muskox,
staples of traditional culture, are
at risk.

—DUDLEY EDMONDSON

SIQIÑIQ, WEARING HER HANDMADE PARKA OF ARCTIC WOLF FUR IN THE
VILLAGE OF NUIQSUT IN THE WESTERN ARCTIC
BY CHAD BROWN

SIQIÑIQ KNEELING IN FRONT OF A GROUP PROTESTING IN FRONT OF THE WHITE HOUSE
COURTESY OF SIQIÑIQ MAUPIN

A TRADITIONAL LANGUAGE INTRODUCTION

I'm going to start with my traditional language introduction. Uvaŋa Siqiñiq, aniruaŋa Utqiaġvigmi. Aakaga Harriet Maupin, appaga Raymond Maupin. Aakaluga Lena Qalhakpak Simmonds, Apaagassi Abe Simmonds Jr. Savaktuŋa Indigenous Wellness Advocate. Panika Ukpiklu Iñuquyuk.

I am Siqiñiq Maupin. I was born in Utqiaġvik, Alaska. My mom is Harriet Maupin, and my dad is Raymond Maupin. My grandmother is the late Lena Qalhakpak Simmonds. My grandfather is

the late Abe Simmonds Jr. I was the cofounder and prior Executive Director of Sovereign Iñupiat for a Living Arctic from 2019–2024. I am now working at the University of Alaska Fairbanks as the Indigenous Wellness Advocate in the Rural Student Services department. I have two daughters, Ukpik and Iñuquyuk. I'm iñupiaq; originally, my family's from Nuiqsut and Utqiaġvik, formerly known as Barrow, where most of them reside.

I was raised primarily in Fairbanks, and I have a home in Fairbanks now. I have a bachelor's degree in Alaska native studies. My concentration is in Alaska native languages with a minor in inūpiaq. I've been learning my language for about five years now, and I hope to continue learning, speaking, and sharing our language—it's one of the most important goals in my life. Along with my two children, I have three dogs, and our little family is the biggest blessing that I have. I've been in recovery for about seven years from alcohol and methamphetamine use, which I think is important to my story and for so many people that have found healing in nature and their culture. We're oftentime dismissed in any important discussions because of our history, but many of us hold vital information on healing, because we were forced to heal. I hope to spread awareness of the healing impact a healthy robust environment has, and the significant negative impact resource extraction can have on the body, spirit, and mind.

I'm a queer, fem person. My pronouns are she/her, they/them. Growing up, I always knew that I was different from the gender and sexuality that I saw around me. It was one of the first, in many times to come, I would know I am different. For too long, the LGBTQ, two-spirit-plus folks in our communities were erased. We have been told that we didn't exist and that we are a new modern problem. Now we're learning that historically we were honored because we knew a different way of thinking. We understood both gender roles, that of provider and caretaker, and we were able to think in a way that others weren't.

RECOGNIZE YOUR OWN STRENGTHS

Many times, people think of ADHD or autism when they think of someone who is neurodivergent. I have PTSD, which is also a form of neurodivergence. So, physically I think differently than the norm. My entire existence goes against the grain.

I've always had a tendency to run late for everything and be a bit disorganized and other things that I used to look at as negative. Eventually, though, I realized that I could think of big-level stuff and do things like create websites in a few hours or get an art piece done in three days because I'd be sleeping on the floor and waking up to work. My brain just works differently. When I meet other neurodivergent people that have low self-esteem because of the world that we live in,

My brain just works differently. When I meet other neurodivergent people that have low self-esteem because of the world that we live in, I love being able to say, you have a unique strength.

I love being able to say, you have a unique strength.

My recovery is also a strength. During recovery, I felt so damaged, so broken that I passionately sought joy, love, and peace in my life, and I have a lot of information to offer others about that journey. I hope to empower others that may be queer or don't fit in. Neurodivergent people bring something unique to the table, and we all are needed in the community.

I also recognize the abuse that can occur with vulnerable populations, like myself and others with mental health issues and traumatic backgrounds. Many vulnerable people find refuge and understanding within the nonprofit world, but much of the time we are exploited. Balance must be brought back using an intersectional lens that doesn't dispose of those who have had a hard past, mental health issues or diagnosis, neurodivergence, or part of the 2SLGBTQ+ community. Our joy and our safety are part of this work to create the world we want to see.

CHILDHOOD MEMORIES OF CHRISTMAS WITH FAMILY

I remember during Christmas, my whole family from the villages would come down to Fairbanks. There were hotels that had rooms with full kitchens, and we would rent out a whole floor. I have 8 aunties, 3 uncles, and over 30 first cousins on my mom's side. We would open all the doors of the rooms and run in and out. My aunties would bring down traditional food in freezers. We would all gather around and cut our native food on cardboard over cutting boards. Many of our traditional foods are eaten frozen and raw, and all us cousins would have tuttu quaq (raw frozen caribou), fish, all parts of the whale, and it is and was medicine for the body and spirit. The kids would try and get muktuk (whale skin and blubber) that was being cut and they'd slap our hands and reprimand us. We were so excited. It was a beautiful thing. Now, my kids and I are the ones who typically bring muktuk and native things to our family gatherings. Now I have to tell the little ones, "No, get out of here," and "You better wash your hands." It's beautiful to see. Those are my favorite childhood memories: spending time with my cousins and eating traditional foods.

ON THE LAND WE THRIVED, WE HAD GOOD LIVES

Learning my language has been one of the top things that I have done to heal generational trauma. Our language was one of the first things they took away from us in boarding schools. Our traditional spirituality was taken away and replaced with this Christian, almost right-wing, white supremacist type of thinking that came up north with white fishermen in the commercial industry. They brought with them the idea of this feared, mean god of punishment that we had to be scared of.

I learned that, in my culture, everything is based on the land and animals. Our traditions sustained us and kept us alive and able to thrive in the harshest environments in the world. Our language has over 50 ways to say ice, and I used to think that was kind of silly, but it's because we needed to know everything about our surroundings. We needed to know exactly what kind of ice, what shape it was in, and how hard the density, so we could do things like find freshwater, build shelters, and travel over it long distances. We still base most of our practices, celebrations, and school around hunting and harvesting seasons. When you talk to elders, they talk about how the land provided and not only helped us survive but thrive. Our people had good lives, even though it was hard. Back then, we didn't have the high rates of things like suicide, depression, and anxiety that

SIQIÑIQ PREPARING TRADITIONAL MEAT WITH HER DAUGHTERS
COURTESY OF SIQIÑIQ MAUPIN

are ravaging our youth today and threatening future generations.

RESOURCE EXTRACTION HAS ALWAYS BEEN AT THE HEART OF COLONIZATION

When I was younger, I worked at my regional corporation. In the 1970s, Indigenous Alaskans were split into 12 regions via the Alaska Native Claims Settlement Act, which you could think of as a series of treaties. Regional corporations were created to be administrators for each region, which inherently goes against Indigenous values. I think we were set up to be where we are today, where monetary value and economics takes precedence over everything and has taken hold of leaders inside of our communities.

Some Arctic Slope Regional Corporation representatives in our community make $10 million a year. These are folks who have come from poverty or what people would consider third-world conditions. No running water, honey buckets for toilets. Now these people have multimillion-dollar homes in Anchorage. They send their kids to Michael Jordan's basketball camp and Space Camp. In our villages, you see those Arctic Slope Regional Corporation representatives and their families get priority for jobs, contracts, and influence on the economy and acceptable opinions about how we make our money. They can do almost whatever they want because many people know they have the power to blacklist those who do not conform.

When I worked for my corporation, I felt privileged because I made a lot of money as an intern. I was making $20 an hour. I was a single mom in addiction, so having flexible hours was the only way I could survive at the time. But I saw a lot of things that felt off. Once, we were pulled into a meeting where they told us we needed to vote "no" on proposition number one, which would tax oil companies. I asked questions about it. Why wouldn't we want to tax the oil companies? Taxes give us money for things that we need in Alaska. They said if we voted no, we would be investing in our people and communities. Voting no was going to feed our children. If we voted for taxing oil companies, those companies were going to have to take their investments to North Dakota because they have less environmental regulations and costs associated with drilling.

While other native interns made books in their languages and visited cultural centers, we were being molded into oil workers and fossil-fuel supporters. They targeted the interns because we were looked at as some of the leaders in our communities. Everything has been infiltrated by major corporations, like BP and ConocoPhillips, that have billions of dollars to contribute to schools and programs.

I've spoken before about the generational trauma introduced by the oil companies. Resource extraction has always been at the heart of colonization, as has violence against Indigenous, Black, Brown, and people of color. Those forces destabilize the economics of community systems, families, and allow outsiders to steal identities and exploit the land without the interference of the original caretakers.

IT'S DELUSIONAL TO THINK WE CAN SUPPORT CAPITALISM AND KEEP OUR CULTURAL TRADITIONS

I worked with the organization Sovereign Iñupiat for a Living Arctic (SILA). We wanted to create a website that shared what we learned at the Frontline Oil and Gas Conference hosted by the Ponca Nation in Ponca City, Oklahoma. It brought together hundreds of activists from communities affected by the oil and gas industry. In our area, over the last couple of years, we have been spearheading the campaign against the Willow Master Development Plan. It's the largest oil and gas project on public lands. The fight has been an enormous challenge, and it has gotten a lot of national and international attention. We were able to bring in an all-Iñupiat delegation to D.C., including two elders who had flown a long way. We were able to tell our stories and explain that our lives are not some romanticized novelty where we run with caribou through pristine places. We're fighting because people are dying

of asthma, respiratory illnesses, and rare cancers. We're doing this because our whales and caribou are in jeopardy. We have been eating those foods for thousands of years, and now we're being told that we may not have access to them in 10–20 years. That is scary for me and so many because we love our land, our animals, and our way of life.

The Arctic is warming at a fast pace, and corporations have so much to gain from telling us everything is fine. They say, "Actually, it's an opportunity for economic growth. Climate change is great! Now, we can have more cruise ships. Now, we can create more economic value by building things like the Willow Project." Corporations are pressuring us to believe the delusion that we can support capitalism and keep our cultural traditions. They say this even though we can see it's not true. In 2023, ConocoPhillips made $11 billion in profit. Clearly, what they have to say is biased.

Arctic Slope Regional Corporation has shareholder meetings once a year, and I attended and started to ask questions. Why in our financial plan do we not have climate

> We have been eating those foods for thousands of years, and now we're being told that we may not have access to them in 10–20 years.

change as a part of our future costs? Why are we not looking at the cost of relocation? Ten years ago, it was predicted Barrow would be relocated in 20 years, and that is coming true. We are seeing entire roadways being washed away. You could see homes in Nome floating down the river. We are on the precipice of having a serious catastrophe. We can see it. We can see black bone marrow in our caribou, which is a sign of starvation. We have begun to see unusual diseases in moose populations. The Arctic, in general, is feeling those changes. We have a very diverse Indigenous population in Alaska, and we're not isolated. What we do in the Arctic Slope is not going to stay there. It's going to go into Gwich'in territory, it's going to go into the Yup'ik and the Athabaskan/Dené people, and the entire Arctic.

DON'T BE AFRAID TO GO AGAINST THE MAJORITY

When you think differently, you can feel isolated. I went through times where it felt like 95% of the Arctic Slope was against the views that I had. I faced a lot of online bashing, and it was kind of a nightmare. I had to get off social media. But respectful elders and other people in the community said things like, "Hey, I actually went through that as a youth too. People ostracized me." That made me feel like I'm getting the support I need, and I have people I can bounce ideas off of. It helps you feel stronger. I really hope that we can change our way of thinking and think of a radically, totally different future.

LEARN MORE

Sovereign Iñupiat for a Living Arctic
silainuat.org

Siqiñiq is a seasoned fighter for environmental justice for the people of the Arctic. A keeper of traditional culture and ways of living in harmony with the land, they have spoken out loudly against big oil companies and the negative affects this industry has on the people and the wildlife that call the Arctic home. I know this work is exhausting, and I applaud all those involved in this fight for environmental justice. —DUDLEY EDMONDSON

DWARF FIREWEED BLOSSOMS ON THE TUNDRA OF ALASKA
BY DUDLEY EDMONDSON

NIKOLA ALEXANDRE

FORESTER AND CO-CREATOR OF SHELTERWOOD COLLECTIVE

CAZADERO, CA

Nikola Alexandre is co-creator of Shelterwood Collective, a community of BIPOC queer folks focused on land stewardship, education, and helping the environment. Land management is a huge piece of what they do, and so is understanding how Indigenous stewardship of the land helps maintain forest health. This is the complete opposite of more-traditional conservationists, like John Muir, who wanted empty land and failed to recognize that humans and nature are one.

—DUDLEY EDMONDSON

NIKOLA (FOREFRONT) WITH MEMBERS OF HIS BLOOD AND CHOSEN FAMILIES; LEFT TO RIGHT, AZZAN QUICK, CLAIRE ALEXANDRE, JOAN LORA, AND JOSÉ BECERRA

BY DUDLEY EDMONDSON

STEWARD OF THE LAND

I identify as a Black, queer forester. I'm based out of Cazadero, California, two hours north of the San Francisco Bay Area. I'm the co-creator and stewardship lead of Shelterwood Collective. It's a 900-acre forest and sanctuary that the Collective founded in 2021. That's where I live, it's where I work, it's where I nurture community. We are an all-queer, all-people-of-color organization. But it's not meant to be exclusively for BIPOC. It's intentionally designed for us, but white folks are also part of the work because we can't get to the changes we advocate for without white folks being true allies and being part of it.

We are a horizontally governed collective, a group of chosen family members, and a formal 501(c)3 nonprofit all at once. We don't believe land, as a living entity, can be owned, but we do recognize the historical importance of Black and Indigenous land security. So, we say we own the title, but not the land itself. We say that we are stewards of this space on behalf of and in relationship with the Indigenous tribe whose lands we are on, which is the Kashia Band of Pomo Indians. We work closely with them to think about, and to model, what Black and Indigenous alliances could look like. In this particular ecological and social climate, we believe in land back efforts, and we advocate for land back as a tool for climate change adaptation and conservation. We also spend a lot of time thinking about what it means to be stolen people on stolen land and to be true allies in each other's struggles, while honoring our own ancestry.

NIKOLA AS A LITTLE BOY ON HIS FATHER'S SHOULDERS IN THE DESERTS OF NEW MEXICO
COURTESY OF NIKOLA ALEXANDRE

CHILDHOOD MEMORIES IN THE NEW MEXICO DESERT

I had a very intimate relationship with nature growing up as a child. Socially, I faced a lot of outward expressions of racism and

THE SHELTERWOOD COLLECTIVE CAMPUS AND THE MAIN EDUCATIONAL BUILDING VISIBLE BEHIND THE TREES
BY DUDLEY EDMONDSON

homophobia, so I spent a lot of time hiking in the woods, running with my dog, camping under the stars. That's where I found solace and community from the lack of community around me. That was accidental. My parents were trying to get away from a lot of oppressive societal systems that were based in capitalism, based in extraction, based in forms of racial expression that objectified my father. That brought them to rural areas, both in the US and in France. We were always coded as outsiders, and then layered on top of that, of course, was the outward expression of my sexuality and of my racial identities. My fondest memories growing up as a kid were running around in the deserts of New Mexico and the mountains of Southern France.

My dad would be behind me with a very small machete that he used to kill the rattlesnakes that I would unknowingly rile up while I was just being a child. But going off into the woods and discovering new places in nature and relating to the trees, the rocks, crying with my dog after experiencing a homophobic comment in the schoolyard, or finding joy while discovering a marmot hole or a little den of flowers, I think that's what helped me see myself as part of the natural world. The land cared for me when I needed it most.

SHELTERWOOD COLLECTIVE, BEING IN RELATIONSHIP WITH LAND AND PEOPLE

Shelterwood was created as a promise, a love letter, an invitation to sanctuary for communities that are feeling deeply down, even if they can't really articulate it, a call to be in relationship with each other and with the rest of nature. Despite all the media that you might see, all the stories that are out there primarily targeting white folks and their relationship to the outdoors, people of color are people of the land, people of the earth, and we can find a lot of solace and healing in one another and our relationship to the land. There was a strong searching for community that motivated me to build Shelterwood, to offer the kinds of spaces that I wish I had when I first started to think about my identity in relationship to the US settler colony and the communities that I was trying to connect to. But it was also informed by all of my undergrad and then graduate career, which was focused on ecology and land stewardship and leveraging the power of nature to fight climate change and help to create a world that is not suffering from climate collapse. What I learned over the years is that the only way for our ecosystems to be healthy is if there are communities of people who are in deep and intimate and generational relationship with those places.

> It's a failure of imagination to not realize that human beings can be more than destructive, but that they can be generative to the land.

IMAGINE A RELATIONSHIP TO LAND THAT TRANSCENDS COMMODIFICATION AND EXTRACTION

The first form of forest degradation that happened in the US was when Indigenous communities were forcibly displaced from the lands they had tended for over 10,000 years. Most people don't think of that, they think of logging as being one of the first forms, the original sin against nature. Indigenous communities played critical roles in keeping the land healthy. The white colonial mentality is how we got here in the first place. It's a failure of imagination to not realize that human beings can be more than destructive, but that they can be generative to the land. Many Indigenous communities have the capacity to imagine a relationship to the land that transcends commodification and extraction. I've met few mainstream conservationist leaders who have the ability to imagine that and the courage to actually act on it within their programs. It's, apparently, a very radical thought. The idea that our national parks should be turned over to communities that are in deep relationship with them, is incredibly radical, and yet it's the only way forward.

Colonial forms of conservation rely on fortress conservation, which does not recognize that people are part of nature. I think science and many of our state and national conservation agencies are slowly catching up to how flawed that modality is. The privatization of nature has evolved into most big institutions only funding conservation that has some sort of financial return. Good examples of this are carbon credits, payments for ecosystem services, and the like. Because of the way our government has been set up, there's the question of where does the money come from to fund these programs? But that just perpetuates this extractive mode of seeing and portraying nature as having to generate income in order to be cared for. If your grandmother needed care, would you first ask her to pay you for it? Now, imagine your grandmother needed care after you were the one who undercooked her chicken.

QUEER AND BLACK AND BROWN BODIES NEED SANCTUARY

When I think of my work, a lot of the choices I made were to be the kind of role model that I wish I'd

A WATERFALL ON BEARPEN CREEK ON THE SHELTERWOOD COLLECTIVE'S 900 ACRES OF LAND
BY DUDLEY EDMONDSON

VOLUNTEERS PULLING INVASIVE FRENCH BROOM FROM THE FOREST FLOOR
BY DUDLEY EDMONDSON

had growing up. That's my guiding star. But to talk specifically about queerness, I did not let myself say yes to the calling of being an environmental steward until I attended a queer people of color retreat in Northern California right after the Pulse Massacre. That's because I didn't see a place for Blackness or queerness, and even less for the intersections of those two things in the environmental world. I attended that retreat, and it was the first time that I was outdoors with 30 other Black and Brown queer bodies. It was a space where the land held this beautiful and transformative moment of grief, and celebration of reconnection.

A lot of folks had never felt safe in the outdoors before. Being able to imagine themselves outside of the confines of heteropatriarchy, of capitalism, of the kinds of violences done to us because of how we are born. For me, that was the first time where I was like, "Oh, this is community coming together. This is healing happening right in front of my eyes." There's a clear need for this, and I want to be a part of whatever work can help create these sanctuary spaces for us. There are not a lot of sanctuary spaces for us that exist outside of nightlife. Nightlife is important to the queer community, but it's not the only kind of space that we need to build with one another. We will continue to claim those spaces, and we need other spaces to call our own and to feel as safe, as the rest of the US deals with things like the "Don't Say Gay" bill and all of the anti-trans legislation. Sanctuary is important for people that have had violent things done to their bodies, to their minds, to their imaginations.

WE HAVE TO LEARN HOW TO BE IN COMMUNITY AGAIN TO SURVIVE CLIMATE CHANGE

Look at the socio-anthropological history of Western colonization, and you'll see there's a very intentional movement to create silos. The nuclear family is venerated across Western society. The individual is venerated. In order to survive climate change, the collapse of our ecologies, we need communities to come together and to tend to the land and to one another. We have to learn how to be in community again. We have to learn how to expand the notion of what is family beyond just that nuclear family definition. Queer folks know how to do that. So many of us have been kicked out of our nuclear families. We've had to reinvent our families. So that practice at Shelterwood has allowed us to create community that transcends those traditional

norms. We invite people to collapse when they come here. A lot of folks carry quite a lot on their shoulders. We want to be a hub for Black, Brown, queer social activists. This is meant to be a space where people can come and collapse and feel what they need to feel, be held by the land, start to nurture whatever relationship they want with the land before going back out and doing the work they do for social change or deciding to stay and do that work here.

CONSERVATION SUSTAINABILITY OUTSIDE THE FORTRESS CONSERVATION MODEL

We're building our model and hope to be able to sustain it by being a collectively run nonprofit. We invite groups that are in better positions to pay for access to the retreat facilities, so that we can sponsor other organizations and we won't have to rely as much on philanthropy, which continues to be a pretty extractive and unsustainable way of doing our operations. We do use philanthropic donations to cover our time and to cover the stewardship of the space. Any group is welcome, but we prioritize Black, Indigenous, and queer groups and provide a series of best practices. We call it our Right Relationship guiding principles, which helps people understand what we encourage them to do and how we encourage them to relate to the land while they are here. We are a retreat center, but we're trying to make it as free and comfortable and as accessible to people who don't have these kinds of nature-based community centers available to them.

I can only be in this work because I'm a cautious optimist. If I didn't think it was meaningful, I couldn't do it. I'm hoping that the creation of these models, their proliferation, their communication, can help shift conservation away from this fortress conservation model and help start conversations around reparations and land back and the return of financial resources that allow these communities to do the work that they need to be doing. It's expensive to tend land, and it can't just be "Okay, we're giving you land." The resources that were generated on our backs as Black and Indigenous people must also be returned so that the work can be carried on. Otherwise, we're just going to be set up for failure. As we think about restoring our forests, as we think about keeping our lands healthy and bringing back a lot of life that has been removed due to colonization and capitalism, we really have to look at bringing back communities. It's an invitation to dream of a climate solution, an ecological solution that relies on communities to help create a pathway for survival through all the ecological and social crises that we're living with.

LEARN MORE

Shelterwood Collective
shelterwoodcollective.org

Nikola is the co-creator of Shelterwood Collective in Cazadero, California. His work there is ground-breaking in that it aims to restore and reconnect Black and Indigenous peoples' cultural and spiritual connections to the land and nature. A critical part of this work is educating them about land stewardship and instilling the values of being in the right relationship with the land. The work is truly radical on many levels, but what if we could all engage the natural world in this way? I think it would make us all much more mentally healthy. —DUDLEY EDMONDSON

NIKOLA SHOWING A FRESHLY PULLED INVASIVE FRENCH BROOM PLANT
BY DUDLEY EDMONDSON

TAMARA LAYDEN

ECOLOGIST

FORT COLLINS, CO

Tamara Layden is a very driven young person. Her life experiences and ancestral South Asian heritage have heavily influenced her viewpoints and approach to science. Combined with a Western education, she is a young female scientist of color who clearly has a bright future ahead of her.

—DUDLEY EDMONDSON

TAMARA STANDS ON THE TUNDRA DURING AN EXPEDITION
INSIDE THE ARCTIC NATIONAL WILDLIFE REFUGE.
BY CHAD BROWN

DRAWN TO CONSERVATION

I am an ecologist currently working on my graduate degree at Colorado State University in Fort Collins. I seek to bridge ecological knowledge and social science for a more holistic view of our environment. I grew up in the Pacific Northwest. My mom was born in India but was adopted at a young age by white missionaries and brought to the United States. Growing up in Oregon, I was in the outdoors all the time. I started volunteering for different organizations like the Audubon Society of Portland and the Xerces Society for Invertebrate Conservation because I was really drawn to conservation.

TAMARA AS A LITTLE GIRL
COURTESY OF TAMARA LAYDEN

I started volunteering at all those places because I was asking the question, what does my world need the most? Where could I fit in and use my skillset? I'm really good at organization and logical thinking, so I leaned into that. I ended up getting this apprenticeship with Audubon. They were bringing in youth of color interested in conservation as part of their mission. After that, I went on volunteering elsewhere. I did return to Audubon to do another apprenticeship, but I was just seeing my supervisors not get supported in efforts to bring in more BIPOC youth. I think I was one of the last apprentices in the conservation department, and then they cut the conservation science component out and focused only on trail stewardship and education. I was like, is this all I can do?

THE CENTER FOR DIVERSITY AND THE ENVIRONMENT

Eventually I went to Oregon State to finish a degree in zoology. I think what really brought it home for me was I took this training with the Center for Diversity and the Environment (CDE). The E-42 program is for emerging leaders with an emphasis on people of color being a majority by the year 2042 or 43. The idea was to help us

be able to lead in a changing world both socially and environmentally.

The program really centered Indigenous folks, especially the fact that people weren't always so destructive toward the environment. In fact, some of the best management has come from Indigenous people living on the land. The separation of people from place is a fallacy we created to basically have dominance over nature and the world. I realized that's why I feel like I don't fit into the science realm. I don't agree with the removal of people from place narrative. Many conservationists believe that narrative, too, separating people from place.

The feeling that humans are too destructive. So I learned through my experience in CDE's training program, and connecting the dots of my past experiences, to know that's not how it needs to be.

I was still getting into that science component through my time managing freshwater ecology labs at Oregon State and Reed College and really enjoying the work, but lingering in the background was how do I bring in this social piece? I almost felt like I had to create something from scratch, which shouldn't be the case.

I started to explore and learn from folks who were doing a bit more of that social component, bringing in Indigenous knowledge, focusing on Indigenous communities and how they're managing the land. So that's where I was trying to shift, shift toward connecting people with place again.

> **I want to help Indigenous communities use ecological and Western science tools, essentially use colonizer tools against the colonizer, to increase Indigenous sovereignty and propagate Indigenous stewardship.**

I want to help Indigenous communities use ecological and Western science tools, essentially use colonizer tools against the colonizer, to increase Indigenous sovereignty and propagate Indigenous stewardship. That's my kind of place in terms of where my background in Western science will go. I think that Indigenous knowledge and traditional ecological knowledge are very valuable in understanding our world.

I remember as a kid, my relationship with nature didn't feel like how science is framed today in terms of "I'm over here, the stuff I'm looking at is outside of me, and I'm staring at it with a clipboard." It was very much with it and within it. I would love to go back to that through my research and conceptualize, what would the science look like that pushes back against assimilation and colonization? What does that world look like?

ENVIRONMENTAL JUSTICE FOR PEOPLE AND URBAN SPACES

I consider myself a conservationist, but one thing we all do poorly is identifying what qualifies as worthy of protecting. These spaces rarely include urban spaces or nature within urban spaces. So what happens there is a lot of disinvestment. Pollution happens in urban areas, particularly for Black and Brown communities, while conservationists create protected areas far away from people or, worse, displace people to create those spaces. You can't say a tree in my backyard is less important than a tree out in Yosemite. I think a fundamental shift is really important here, and if it can happen in conservation, there's a lot of power that comes with that.

I've been reflecting on place-based knowledge versus Western science a lot recently, as well as in my research, because something that I have seen in my mom is this really ingrained connection to land and gardening. It's just something that she knows really well, and I haven't been able to fully understand it, and I think it's because of my Western science background. She'll come over and she'll just be telling me what my garden needs, and jokingly, I'm like, where's your data? Where does this come from? It's

just a different way of knowing, a way that's really deeply connected with the land and the place that she's in. I don't have that same connection that my mom has. I know it's related to this Eurocentric way of knowing that aims to separate emotion from your work. It aims to separate intuition from your work. I think that there's been this tremendous loss because of that. Yeah, it just makes me think of how much that ways of knowing, the Eurocentric way of knowing, has honestly done a disservice to science and what we claim the values to be. It's that dominion over nature, that Eurocentric concept, coming through. This is tragic, honestly. I think about that a lot in my work and how can I shift my own language and my own process.

INCREASE ETHNIC DIVERSITY IN SCIENCE BY EXPANDING ITS DEFINITION

I think one way of making careers like my own more culturally diverse would be to think about how we define science. There are some people of color like me who are in the Western science field. I think further legitimizing and validating Indigenous knowledges—that is, place-based knowledge that we've undervalued in Western science—and the knowledges generated over generations from enslaved Africans, where Black folks were working with the land, on the land, could help. We just don't see that in our practices of agriculture. We don't talk about that knowledge. I think legitimizing those ways of knowing as science could help expand ethnic diversity in environmental science.

CONSERVATION AND THE FUTURE

Society has been faced with a racial reckoning of late, and we can argue about whether or not it did anything with it, but at least people had to have that conversation. I think conservationists also need to be having these conversations and thinking about it. I think change can come from grassroots movements, but it's definitely hard for those to exist and thrive if the conservation model isn't challenged. Finding and maintaining hope these days is truly tough, but it wasn't until I started looking into examples of subsistence and ancestral stewardship that I really started to see a future beyond the imperial capitalist age we are so often consumed by. These examples, coupled with powerful and radical youth, especially youth and women of color, really inspire me. Young people and my peers are becoming frustrated, and that is not necessarily a bad thing. I live for the day we empower each other to say enough is enough and take our world back.

LEARN MORE

tamarajlayden.wixsite.com/ecology

The Center for Diversity and the Environment
cdeinspires.org

Indigenous Land & Data Stewards Lab
indigenouslandstewards.org/who-we-are

Tamara is a bright, young, Brown scientist who sees the scientific world differently. She is concerned about how Western science influences place-based knowledge and that of Indigenous people. Her work centers on helping Indigenous people use Western science to uplift their cultural ways of being. If those in conservation could only let Indigenous nations lead, I think we would finally gain some ground in repairing the harm we've done to our planet. —DUDLEY EDMONDSON

TAMARA SETTING UP A TRAIL CAMERA AS PART OF A RESEARCH PROJECT
COURTESY OF TAMARA LAYDEN

RICKY DEFOE

TRIBAL ELDER AND OJIBWE LANGUAGE SPECIALIST

CLOQUET, MN

Ricky DeFoe is an Ojibwe elder and always has these amazing points of wisdom. Whenever I spend time with him, we have these great conversations and learn about the people, the environment, and the many connections we all have.

—DUDLEY EDMONDSON

RICKY STANDS ON THE BANKS OF THE ST. LOUIS RIVER,
DOWNSTREAM FROM THE FOND DU LAC INDIAN RESERVATION.
BY DUDLEY EDMONDSON

RICKY WITH HIS GRANDSONS, SIGHTSEEING DOWN IN THE DULUTH-SUPERIOR HARBOR IN DULUTH, MINNESOTA
COURTESY OF RICKY DEFOE

MY LIFE, CONTINUING THE TRADITIONS OF THE OJIBWE PEOPLE

I was a journeyman ironworker, but now I am retired. I switched careers from doing physical work to mental work, and I'm part of the culture and language revitalization efforts here on the Fond du Lac Reservation. I also serve on the Water Legacy Board of Directors in Duluth, Minnesota, as well as the Antiracism Study Dialogue Circle Board of Directors in Prior Lake, Minnesota.

My job title is Ojibwe Language Specialist, meaning I specialized in the language of the Native American people of this region. I am building a curriculum to improve our language skills. Nowadays, it's a written language, but in the past, it was an oral language, Anishinaabe. Our language is rooted in space- and place-based knowledge. We can't go across the seas to Europe, France, or wherever to get our language. It's here. We work with the linguists among our own peoples. I have groups come in to do teaching moments around various concepts. We also utilize talking circles where we exercise our listening skills and practice ceremony at the

Arrowhead Juvenile Center, and Northeast Regional Corrections Center, here in northern Minnesota. We also go to the hospitals and do our part in praying for those that are sick, the homeless, and things of that nature. We reach out as a part of the community and bring some healing with us.

MY CHILDHOOD, SURROUNDED BY LOVE AND KNOWLEDGE

I was born in 1959, in Cloquet, Minnesota, on the Fond du Lac Reservation. After the Relocation Acts of 1953, my family relocated to Duluth, Minnesota. Other families went to Minneapolis, some went to Cleveland, some went out west. The stories were that they most often ended up in the inner city, in depressed areas, struggling while away from their homelands.

We had a huge extended family versed in Ojibwe culture and Indigenous culture, which includes much more than just a nuclear family. Many of our relatives stayed in the house. In the beginning here, we just had small tar paper shacks and little old shelters, rather than homes. Later, there was the construction of a housing complex that had running water and electricity.

When I was younger, you could always feel the love because you were right there with the elders and their teachings and their soft words, kindness, smiles, and hugs. There was so much love from the elders to the children. The things I remember most were taking walks with my grandfather and picking things like berries. My uncles would take us to out into the woods to hunt rabbits and deer, with the bigger rifles. Even though we were raised on the reservation, before moving to Duluth, we still practiced some of the holiday traditions of mainstream America, like Christmas. The giving from elders and the way they took care of us was really something to behold. Because we were living in the same home, everybody was contributing. My uncle would go away to work as a

RICKY IS AN ELDER AND A PIPE BEARER FOR THE FOND DU LAC BAND OF LAKE SUPERIOR CHIPPEWA.
BY DUDLEY EDMONDSON

lumberjack and save his money up and come back and disperse it. The moments when we reconnected were happy times. It was much more of an inclusive atmosphere than it is today. Everything now is more selfish, and nobody wants to help each other. Individualism creates extended hardships. I see the importance of being in a communal mindset because if we lift each other spiritually and emotionally, mentally and physically, then we are happier and healthier. I think the biggest piece is to revitalize that communal way of seeing and being in relationship with others.

THE MAN-MADE WORLD IS AN ILLUSION

We are born into this world that we think is a real world. But then as we talk to the wise ones, the thinkers in our communities, we realize that it's kind of an illusion. People talk about manmade economic exchanges and that's supposed to bring happiness. If we have a big bank account and the stock exchange is up, then everybody's happy, but that's just a falsehood. And so we strive to become rich and live the American dream, but we get sick along the way. So how do we return to the real world?

There are different forms of knowledge. We understand that intergenerational knowledge is very, very important. There is also dream-state knowledge that goes beyond the physical into the spiritual. Then we have ceremonial knowledge and contemporary knowledge. Oftentimes contemporary knowledge is the one that's misleading, and that's the world we're involved in now. We have to recenter ourselves into this world, into this cosmos, and realize that what we're in is an illusion, because nature and culture are the actual realities.

Knowing that we're all colonized is a good starting point. White people have to understand they, too, have been colonized. We all need to understand that we've been raced, we've been gendered, and we've been colonized, all of us. Then from there, we can start to learn some new things about worldviews and perceptions, understandings of humanness, Anishinaabe, original peoples. Understandings that pre-date all other human structured expressions, including all of the religions of the world, and the social and political orders of the world. The problem with mainstream America and the Eurocentric way of seeing the world is that so-called modern peoples have been influenced by a paradigm that involves dominion over all

THE ST. LOUIS RIVER DECORATED BY BEAUTIFUL FALL FOILAGE DOWNSTREAM FROM THE FOND DU LAC INDIAN RESERVATION
BY DUDLEY EDMONDSON

THE FOND DU LAC RESERVATION CULTURAL LANGUAGE LEARNING CENTER
BY DUDLEY EDMONDSON

things. We hear a lot about that. I've heard about that in the Catholic Church and other places. In fact, the church papal bulls declared war on all non-Christianized peoples. They said that people of color were subhuman and that we could not own land, but that we could be a part of the land. We are just like the animals and the plants. Our way of seeing the world is all things have rights, the animals have rights, that we have rights. The fliers, the walkers, the swimmers, the crawlers, the growing plants, and all others have rights.

The interdependence of all things is what makes up a true democracy.

INDUSTRY IS A PROVEN POLLUTER

The extractive industry is a proven polluter. The onslaught of mining companies like Polymet and Twin Metals and the folks coming from international corporations are a huge threat. There are millions of gallons of "treated" water being dumped into the so-called St. Louis River here in Northern Minnesota. It's the largest tributary to the estuary and then into

Lake Superior. These companies are pushing forward with trying to build bigger dams so they can hold more contaminants, sludge, and other waste materials. I worked at Potlatch Corporation as a modernization ironworker. We'd be up on the iron while they were discharging pollution into the river, and we could see the river and air change colors; sometimes it was yellow, sometimes it was green, but then the monitor alarms would go off and they'd bring us off the structure. All the construction workers would go into the parking lot and we would meet out there and get a head count. We'd be out there for about a half hour, and then they'd give the all-clear sign to go back to work, probably because they'd got done dumping all that stuff into the river.

The shipping industry is also a problem. They dump the bad oil that they have in the sumps directly into the oceans. The fuel that they burn is one of the dirtiest fuels on the planet. They're bringing it right here into the twin ports of Duluth, Minnesota, and Superior, Wisconsin. There's lots of different industries contributing to the pollution. The state values iron ore, copper, nickel, and other metals more than they value the watersheds, the plants, and the

animals. Their values are not in line with what they're saying, so there's always that cognitive dissonance.

RECOGNIZING THE KNOWLEDGE OF INDIGENOUS PEOPLES

Sometimes in environmental justice, there is much talk about reconciliation and coming together. People want to say "All right, now we're friends again. We had a dispute where we stole your land and killed most of your people, but let's reconcile. We don't want to talk about reparations, but we want to reconcile." The wise ones say that without justice, how can you have reconciliation? Justice means action. It's something that you begin to feel when actual things take place, not just talk. To have environmental justice, we've got to have rights of nature, the rights of animals, the rights of trees, the rights of rivers, which is a whole new perspective. That includes recognizing, understanding, and legitimizing traditional experience, traditional knowledge. White America and white supremacists fear that acknowledging the inherent wisdom of traditional peoples will create a threat to their very existence. The knowledge of Black America, Indigenous America, and

> The wise ones say that without justice, how can you have reconciliation? Justice means action.

other nations threatens the way they perceive themselves.

There is a fear that drives the decision-makers in a white-led nation. Believing in the wisdom of other cultures is counterintuitive to the exceptionalism that white Americans have and the way they see themselves. Once they realize that, then I think the activism of Indigenous peoples and brothers and sisters of color, and the way we see the world and do things will be accepted. The climate and Mother Nature will right herself. We have an opportunity to be a part of that righting, to change this upside-down world right-side up. We then must ask ourselves, what abilities do we have? What responsibility do we have? What obligations do we have? And those things we must answer in our quiet times with ourselves.

LEARN MORE

Water Legacy
waterlegacy.org

Antiracism Study Dialogue Circles
asdicircle.org

Ricky is a longtime friend and someone I've always looked up to as a mentor. Ricky is a Tribal Elder on the Nagaajiwanaang aka Fond du Lac Reservation near Cloquet, Minnesota. He is also a well-respected member of the Duluth, Minnesota, and Superior, Wisconsin, area and is often consulted by city officials and community leaders on tribal matters. —DUDLEY EDMONDSON

RICKY HOLDING A SACRED CEREMONIAL PIPE
BY DUDLEY EDMONDSON

JASON HALL

BIOCHEMIST AND FOUNDER OF IN COLOR BIRDING

PHILADELPHIA, PA

Jason Hall is a birder who has a mission to get people of color into the outdoors, for the mental and physical benefits, but also because birding is about having fun. His unique approach to birding is something I think we need. In traditional birding groups, there's always this posturing and competition and an unnecessary level of stress. There's no need to get everything right. If you misidentify a bird, we can all learn together. Birding is about having fun and less about a sudden pop quiz.

—DUDLEY EDMONDSON

JASON BIRDING AT VALLEY FORGE HISTORIC PARK
NEAR PHILADELPHIA
BY DUDLEY EDMONDSON

MY LIFE, MY WORK

I'm the founder of the In Color Birding Club here in Philadelphia. Our group focuses on bringing BIPOC birders into the birding community and creating a collective of birders of all ethnic and cultural backgrounds.

I am also Director of Vaccines in a group that manages reagents used for vaccine quality control. I've spent the better part of the last 20 years in vaccine development and commercialization.

JASON (CENTER) AT CHRISTMAS NEXT TO HIS BROTHERS NATHAN (LEFT) AND ALEX
COURTESY OF JASON HALL

MY CURIOSITY ABOUT NATURE STARTED WHEN I WAS A KID, GROWING UP IN CALIFORNIA

I cannot remember a time that I was not in nature. I credit my parents for that. My father grew up in rural Mississippi and my mother in southeastern Idaho. Both of my parents were from families that relied on agriculture or farming to some degree, with a consistent connection with the land. I remember spending a lot of time in my backyard in Southern California. It was a relatively large backyard and had a bunch of fruit trees. When the Santa Ana winds would blow, these huge white pines in the front yard would sway around in the strong winds. Every time, I remember thinking those pines are going to snap at any given moment. The next morning, I would come out and be astonished that they had not snapped. I remember thinking to myself while walking underneath them, "What is underneath the ground that is holding you in there like that?" I spent much of my formative years as a child in that yard exploring the environment and my place in it. That was until my parents divorced when I was 8 years old.

I moved to Mississippi with my father. He had a very small house. It was me; my two brothers Nathan

A LARGE FLOCK OF EUROPEAN STARLINGS SETTLES TO THE GROUND IN A GRASSY FIELD.
BY BILDAGENTUR ZOONAR GMBH/SHUTTERSTOCK

and Alex; my sister, Octavia; her daughter, Ryan; my dad; and his wife, Wilhelmina, all packed in a two-bedroom house. It just wasn't big enough, so my dad and his wife worked hard to buy a bigger house and particularly one with another massive backyard. Instead of white pines, it had massive pecan trees and a huge forest of bamboo that obviously was not supposed to be there. I remember watching European starlings come in to roost in the bamboo each fall. They had a flight pattern where they would kind of split up into four trees, and then each tree of birds would take turns swooping into the bamboo. That would continue until the bamboo was just hanging over sideways

because it was so heavy, full of birds roosting for the night. My dad had this big greenhouse, and I remember finding dead starlings near that greenhouse. They probably collided with the glass. I can still almost feel them in my hands, picking them up in that thick Southern bluegrass. It looked really beautiful. It sparked a conversation between me and my dad, who is heavily religious, and I am heavily skeptical. We would have these great discussions that didn't always result in any agreement. But seeing my curiosity about these birds inspired him to tell me stories about his childhood in rural Mississippi. Seeing me, the explorer in the backyard, gave him reason to have a conversation with me about his own childhood experience with science and curiosity, and that brings back great memories. My life, like many folks, has pockets of trauma and things that we probably want to forget. But they are buffered in many cases by my relationship with nature and how that manifested in my life and with the people that I love.

> My life, like many folks, has pockets of trauma and things that we probably want to forget. But they are buffered in many cases by my relationship with nature and how that manifested in my life and with the people that I love.

MY MOM WAS THE KIND OF PERSON THAT TOOK IN EVERY CREATURE IN NEED

My father tells this story about how we had a baby squirrel that was abandoned. My father did not have the same, let's say, sympathies with nature that I did, so he had my older sister throw it in the trash. He said he looked out 20 minutes later and he saw my feet dangling out of the dumpster because I had jumped in the trash to get the squirrel. He said that was when he realized the connection that I was going to have with nature. I didn't hear that story until I was probably well into my teenage years. That story itself became a healing story between me and my father because he tells that story all the time. When he first met my wife, he told her that story. When he came to my college graduation and met my professors, he told that story. That story happened in the same backyard that my parents would argue in after they got a divorce. Even with all that happening around that space, the thing I remember most is how proud my father was that I jumped into a dumpster to get that squirrel. After the fact, it's now become healing. I have a love of all living things, and that extends to my father, who I felt like made a lot of mistakes when I was growing up, but I was able to forgive him. I wouldn't have been able to move my heart in that direction without a mother who ensured my approach was always an empathetic one first. My mom was the kind of person that took in every creature in need, whether animal or human. She was, and continues to be, *that* house in the neighborhood.

MY LOVE OF NATURE WAS A GIFT MY PARENTS MADE SURE I RECEIVED

I knew at age 13 that I wanted to be a scientist. My mom took me to see *Jurassic Park* and they were, like, putting frog DNA in eggs, and I was like, that's it, you don't have to tell me nothing else. My mother encouraged all my forays into science and worked hard so that I could focus on the things I loved during and after school each day. By my senior year, I had taken every science course required to graduate, plus some. Because I couldn't get enough, I took an Environmental Science course for extra credit. Mr. DiCenna took us outside every morning at 8 a.m. and gave us a notepad and binoculars. He said, "Okay, this semester each of you are going to pick a 20-yard-by 20-yard space and you're going to document the birds and the trees in that space. One of the first things I did was sketch this tufted titmouse that was just hopping around on this tree in front of me. Looking at it, I felt the same things I did walking under the white pine trees in my front yard in California as a kid. This sense of wonder and renewal because something new had entered my brain. If I had to impart anything to my kids, that's probably what it would be. It's just an appreciation of love for nature in whatever way it comes.

Having parents that support your love for something is a very powerful thing. I wouldn't be here without both my mother and father contributing to my pursuit of these passions. Nature essentially represents this gift that they made sure that I was going to receive.

A TUFTED TITMOUSE PERCHED ON A BRANCH; THIS WAS THE BIRD JASON STUDIED DURING AN EXTRA CREDIT ENVIRONMENTAL SCIENCE COURSE IN HIGH SCHOOL.
BY DUDLEY EDMONDSON

BIRDING FOR ALL

I had the idea for starting In Color Birding Club after seeing the tack-sharp focus of people that were 10 or 15 years younger than me who put on Black Birders Week.

I felt there was something that I could do, I just didn't know what. It was COVID, so everything was online. I was just trying to soak up as much as I could around this perspective to see what are other folks doing in their spaces. Where is the need?

Here in Philadelphia there's a long history of birding, but none of the organizations had anything that

I would call a real solid anchor on reaching out to the BIPOC community for birding. And at the same time, I was seeing all these stories about birding being on the decline, because it's all old white men. How do we get younger people into this? And I was sitting there thinking, they wouldn't have such a tough time if they stopped limiting themselves to the same circles and acting the way they do.

WE BIRD DIFFERENTLY BECAUSE THE CHICKADEE DOESN'T CARE IF YOU GIVE ME A HIGH FIVE

The mission of the club is really to open the space to BIPOC birders, and center BIPOC birders. But we're also open to allies because the end goal for us is not to end up with a Black-only birding community or a BIPOC-only birding community. The idea there is to break the systemic structure around birding that was built around white supremacy and all the things that have been ingrained into the different facets of our society. Essentially, just make all that crumble and create something new out of that rubble. What I saw starting to happen at our outings was people saying, "I've never been birding before. I just wanted to get outside." What I found was a lot of people in their 20s and early 30s who at those ages typically have a lot of other options

of things to do, but the pandemic had shut those down. By having our events marketed to communities of color, it presented something new for them, just in terms of how we were marketing it. We were saying, "Hey, no experience needed. We have binoculars if you need them, just show up here."

I found that birding became the perfect activity for those people, as they could have some community and get some exercise outside, without the COVID risk. If this was the tried-and-true, say lily-white birding activities, I think it would have lasted three weeks trying to market in those BIPOC spaces. You just wouldn't have had the kind of community engagement that I think communities of color value. By getting people out there and saying, "Hey, if you see something and it moves, you laugh, cry, high five, dance, whatever you want to do." That was a very different activity than a lot of people expected. They were probably coming in perceiving it as a very quiet activity, walking through the forest and observing. But I was like, yo, if you see the chickadee and the chickadee makes you happy, give me a high five. The chickadee doesn't care if you give me a high five. The chickadee is surviving all kinds of more dangerous things than you and I giving high

JASON TAKES A SELFIE WITH A GROUP OF IN COLOR BIRDING PARTICIPANTS AT FDR PARK IN PHILADELPHIA.
COURTESY OF JASON HALL

A GREAT BLUE HERON STANDS AT THE WATER'S EDGE HUNTING FISH AND FROGS.
BY DUDLEY EDMONDSON

fives. Our club produced a different birding experience, yes, but I haven't had an outing yet where I felt like people left in a worse state than when they came.

PROVIDING MEANINGFUL EXPERIENCES IN NATURE FOR OTHERS IS REWARDING

I think it's working. We've had a lot of the historic birding organizations around Philadelphia reach out to us and ask us to come help participate in things, or they want to come to our walks, or provide monetary sponsorships for different things, particularly our bus program. We took some students out from a local high school in West Philly to bird at a local park. The kids spent probably 30 minutes of that walk just watching this great blue heron stalk frogs in this pond. They have a very special teacher, and he said, "You know, a couple of these kids were having a really bad day. I can tell you, they're the two kids back there with the biggest smiles on their faces right now." That's the kind of stuff that makes you just want to break down and cry, you know. I may never see those kids again and that's fine. But they have that memory now and the thought of I could go do that again if I want, because that space was open to them in a way that was inviting and welcoming. When you squeeze life, that's the stuff that tastes the best. It's knowing that you're, at least for a short period of time, providing some happiness and joy for someone to enter into nature that maybe would not have had that opportunity otherwise.

TACIT KNOWLEDGE VS. SCIENTIFIC KNOWLEDGE

I gave a talk on the history of Black and Brown birders in a place called the Awbury Arboretum in north Philadelphia. I was particularly interested about birders of African descent. In my mind I was thinking, "Okay, let's rewind before European ships landed on the West African coast and started snatching folks. What were those communities like even during colonial occupation? What was going on there?" We have a good history of the Indigenous folks in the Americas and about how they perceive the landscape, how they understood the other species in the landscape, and how they worked together. Native American history shows you very clearly that they understood that wolves had to be here to balance out the population of other creatures. I wondered, what was happening in Africa when it comes to birds? My presentation on that went back to this tried-and-true thing that we

know here of white men trying to put names on stuff. I say that with a slight bit of humor, but it's true.

I found this book, it's called *Birders of Africa*, written by a Brown University history professor, Nancy J. Jacobs. It talks about how these colonial explorers relied on the tacit knowledge of these Indigenous folks to really understand what these birds do. You had European explorers trying to classify these species in these exotic landscapes with no idea where to start. One of the stories was about a bird called a Honey Guide. From the name Honey Guide, it's clear that a colonial explorer obviously gave the bird that name. Well, the story really struck me because this Indigenous tribe member was talking to this explorer about the bird, saying, when they make this noise, you can follow them and they will take you to a beehive and you can extract some honey, but you have to give them a piece of the honey. The lore within the culture was, if you didn't do that, the next time they were going to make the same noise, but this time lead you to a hungry lion.

I was particularly interested about birders of African descent. In my mind I was thinking, "Okay, let's rewind before European ships landed on the West African coast and started snatching folks. What were those communities like even during colonial occupation? What was going on there?"

Obviously, that's just myth, but the lesson was there that you had to have some gratitude and give-and-take with this species.

EUROPEAN EXPLORERS OVER-SIMPLIFIED NATURE, LEAVING OUT THE CULTURE AND RELA-TIONSHIPS BETWEEN HUMANS AND CREATURES

My talk about the history of Black and Brown birders became a discussion about this split between learned knowledge of a local people's understanding of the environment that has extreme value, versus what we know today as ornithology, which is very rigid and scientifically based and all about classifications and less around behaviors. What the European explorers left out was all this lore and learning that the tribes had given them in order to explain what the birds do. They took that information and flipped it into science, leaving behind all the history, the culture, the relationships between those humans and those creatures. This became the point of the talk, understanding the ancestral lands that we're on that were stolen from those Native American tribes and everything that means about how they managed this space.

Our conservation efforts need to be considering how the Indigenous folks of this land were doing that before Europeans arrived. Another lesson is, as Black and Brown folks, we don't have to approach something like birding with just this pure scientific, rigorous mindset. Let's approach it how our ancestors in Africa and the Americas may have approached it prior to being interfered with by colonial forces. Maybe we can kind of reclaim some of that path that we may have been on ancestrally.

RACISM IN THE OUTDOORS

I've had things where people won't even look me in the eye. They won't say hello. You know, those things I got used to over many years and they still happen to this day. I've

---

AUTHOR'S NOTE: Christian Cooper was a Black Central Park birder who encountered a hostile and racist white dog walker named Amy Cooper. When Christian asked her to leash her dog in an area of the park clearly designated as an area where dogs were to remain on leash, she refused. She instead threatened to call the police, and lied, saying that she was being violently threatened by a Black man. She was fully aware that police presence would almost certainly result in Christian being arrested or possibly even killed. Amy Cooper clearly understood her white privilege in choosing to use the police as a weapon against a Black man.

pulled into parking lots, full of birders, and I'm the only person of color. I get out there and I remember, one time, this guy looked me dead in the eye, and he looked back at his wife and said, "Make sure you lock the car." I've had people on the trail ask me what I'm doing there, am I sure that I'm a birder? That's one of my favorites. Ask me, are you sure you're looking at birds? I'm in the middle of the forest, what else would I be doing, bro? I got these binoculars and this big-ass camera, what do you think I am doing?

Those things have happened many times, but I've never had anything as egregious as what Christian Cooper went through in Central Park. I think every person of color knew what that was like. I think had it been something where she was just really ignorant, which happens to a lot of us all the time, it wouldn't have made the news, but using the police as a weapon to obviously try to cause Christian harm was the part that stuck people directly in the chest when they saw it.

At this stage, part of the bird club and changing the culture is trying to make sure some of that stuff doesn't happen to people as much anymore. The Civil Rights movement was not a start-and-end activity. The fact that I had to start a bird club to make sure people of color were getting out into nature says there's still work to do, so I have to take the baton now, and I have to do something with it.

LEARN MORE

In Color Birding
incolorbirding.org

Jason Hall is the founder of the In Color Birding Club. The mission of the club is to open the birding and outdoors space to BIPOC birders. He's also Director of Vaccines at Merck, where he ensures that the quality-control materials used to test and verify the efficacy, safety, and potency of vaccines are managed in a way that provide assurance of the ability of the vaccine to work clinically. ▬DUDLEY EDMONDSON

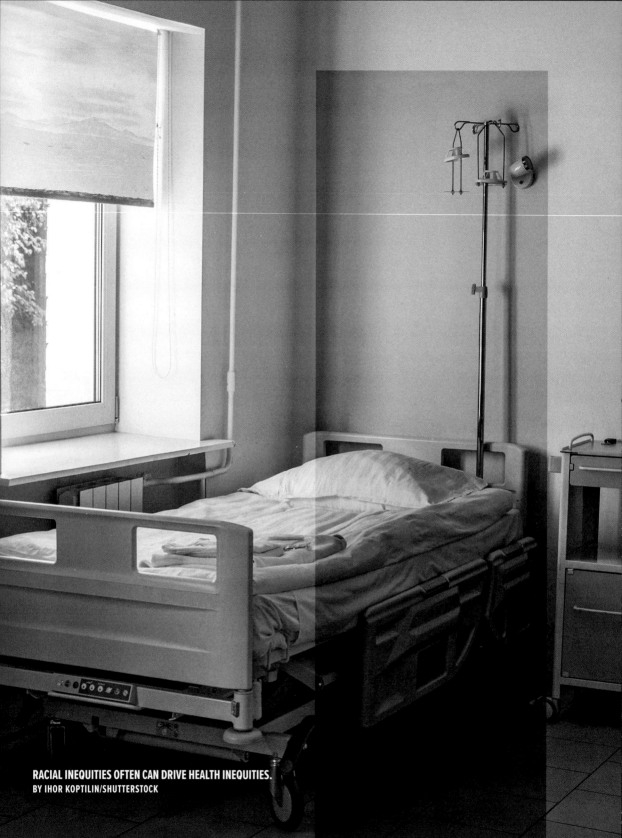

RACIAL INEQUITIES OFTEN CAN DRIVE HEALTH INEQUITIES.
BY IHOR KOPTILIN/SHUTTERSTOCK

WHAT?!

"Social determinants of health, such as poverty, unequal access to health care, lack of education, stigma, and racism, are underlying, contributing factors of health inequities."

—CENTERS FOR DISEASE CONTROL, 2023

ASHANEE KOTTAGE

ENVIRONMENTAL JUSTICE STORYTELLER

WASHINGTON, D.C.

Ashanee Kottage's work involves a fusion of theater, art, and storytelling to convey conservation messages. It's a unique, useful, and ingenious way to communicate science, far better than the typical stuffy "scientist" we're all familiar with. It's genius and her perspective needs to be included.

—DUDLEY EDMONDSON

ASHANEE, AN ENVIRONMENTAL JUSTICE STORYTELLER IN D.C.
BY DUDLEY EDMONDSON

INTRODUCTIONS

If I was directing a movie about my life, what would the three film sets be? The first would be a tropical island by the beach. Being from the tropics near the equator is important to me. Our people had to learn to coexist with flora, fauna, and fresh food that they could see grow. There's an unmatched sense of hospitality on islands, particularly Sri Lanka. I'm proud of that, and all the rich cultural heritage.

ASHANEE AS A LITTLE GIRL AT KANNIYA HOT SPRINGS IN TRINCOMALEE, SRI LANKA
COURTESY OF ASHANEE KOTTAGE

The second would be my high school basketball court. I've always been a big team sports gal and am incredibly collaborative. It's where I learned how to be a leader, how to be a team member, and how to work in a group. So, teams and groups are important to me.

Third would be my mom's shrine room. I was lucky to grow up in a religiously pluralistic background. We are a Buddhist family, and Sri Lanka is a majority-Buddhist country, but that comes with all the problems that you see anywhere with a majority anybody, anything. My mom was creative, introducing us to all religions. In the middle of her shrine room was a massive statue of Lord Buddha and then Hindu deities all around, a little statue of Jesus and Mother Mary, and texts from the Koran.

In my spiritual practice, I embody, worship, pray to, and honor all religions and deities. That faith drives what I do and affects my work as it pertains to nature because I think all these gods are in nature: a leaf is god, and an elephant is god.

CHILDHOOD GROWING UP IN SRI LANKA

My connection with animals is deep and has been ever since I was a child. In Sri Lanka, we have stray dogs everywhere, and they are their own breed because they've existed on the streets for centuries now. I was obsessed with them. I was always on the streets petting and playing with any animals I could find. I was lucky to see elephants roaming temples, peacocks flying around, and dogs chasing squirrels. I wanted to be a part of the game, which also relates to my faith.

When I was growing up, I would be visited by Lord Ganesha in my dreams. Lord Ganesha is half elephant, half person, and he's the patron god of poetry, of the sciences, of reading, and the remover of obstacles. He's just a playful dude. That's what I love about Global Majority deities—they have a lot of complexity and are very multidimensional and relatable. He would visit me in my dreams in elephant form, so I connected with all elephants and animals. He was kind of assigned to me as my patron god because I was academically driven, and he was the god of studies and books and stuff.

I'D ALWAYS KNOWN THAT EDUCATION WAS THE ONLY THING SOMEBODY COULDN'T TAKE FROM ME

I went to a British international school in Sri Lanka, which serves as a pipeline to British colleges. I was enamored by the liberal arts education that the US offered because in the UK you either pick law, medicine, engineering, or whatever. You don't really get to take electives around it. I knew that I wanted to explore. I used to think it was a problem because everybody else seemed to pick something and know that's exactly what they wanted to do. But I was interested in the alchemy of things. I want to do history and math, put them together and see how it goes. I want to do performance art and science, put them together and see how it goes.

But I was interested in the alchemy of things. I want to do history and math, put them together and see how it goes.

When I came to the US in 2018, I was deciding between UC Berkeley and Georgetown. My dad didn't allow me to go to college. He's pretty conservative, pretty traditional in that way. I refused. I had always known that education is the only thing that somebody couldn't take from me. So, I fought, and I was lucky to have my brother as an ally. He's four years older than me, and the opposite, in that he never wanted to go to school but was sent off to college because he "deserved" to get a degree. He said to my dad, "If you don't let her do it, she's going to find a way to do it, and you don't get to be a part of her story anymore." He said, "You get to decide where you fit into her story. Do you want to be a villain? Do you want to be a supporter?" My dad finally came around. So that's how I ended up at Georgetown.

LEARNING TO BE A MULTI-HYPHENATE WOMAN

In the Georgetown School of Foreign Service, they have proseminars that you have to take your freshman fall. It's 15 students and a niche specialized topic, so you develop a close relationship with the professor and 14 other students. I thought, "Oh, I'm never going to take any environmental related classes, so let me poke my head in and see what's up." I took a class called Deeper Sources of Environmental Policy. It was with Dr. Clare Fieseler, who is a Georgetown alum. She was a *National Geographic* explorer, a filmmaker, and a marine biologist. She showed me that you could be a multi-hyphenate, that you could be a photographer, scientist, woman, mother, and professor. She combined a lot of identities that I hadn't really seen combined. That steered me toward environmental studies. In the beginning, I faced a lot of opposition from professors and peers, who didn't realize that you have to follow the globe and

that scientific knowledge comes from around the world and with cultures that are not Eurocentric.

USING PERFORMANCE ART AND STORYTELLING TO COMMUNICATE SCIENCE

I landed on storytelling because I had always been an actor, dancer, and performer, but it was always a hobby. It's a part of our culture. Music and dance are such a big part of Global Majority culture. They were never something I took seriously in myself. My second year in college, a show came to Georgetown called *On the Lawn* that was put on by LubDub Theatre Company, which was founded by Georgetown alums. They put on a show about the consequences of the American lawn on climate change. I auditioned and was like, "Wow, this is so cool," because it was an eco-art installation. It was a physical, devised theater piece, and I was getting class credit for it. I was researching, writing, acting, dancing, and trying a lot of combinations that I never had before but was being allowed to do in the same place at the same time.

Through that play, I realized how effective performance and storytelling are in communicating the things in academia that don't reach anybody because who has the time,

ASHANEE'S PERFORMANCE AT COP27, A UNITED NATIONS CLIMATE CHANGE CONFERENCE HELD IN EL SHEIKH, EGYPT
BY CRAIG GIBSON

ability, and access to read journal articles and long scientific reports and can understand data the way it's presented in Western academia? I realized that entertainment and stories are such a good way to access people's consciousness.

I am constantly drawn back to theater because of how collaborative it is, especially devised theater, because you are not given a script; you research, interview, and create a script with your collaborators. You'll tell a story and then they'll tell a story, and then you'll see what words your stories share. Then you have a whole new vocabulary, and then somebody else can use those words to tell a story. That weaving

is more attractive, meaningful, and enriching to me than a one-sided form of communication.

I see so much art in science and so much science in the inquisitive, interrogating methods in art. When I'm creating, writing, and devising, there is a science to it. And there's such beautiful creativity and artistic integrity in the work that comes out of traditional science labs.

I see the connections in the methods, and I borrow from both in my practice. The connection is also about accessibility and to help science become embodied, to take data, graphs, statistics, and numbers, and help people really feel it in their bodies. I use the art to serve the science, so that when you are consuming some form of information, you are witnessing it in your heart before you witness it in your mind.

IT SHOULDN'T BE A LUXURY TO SEE NATURE

The scholar Mays Imad talks about unspoken agreements in STEM and data analysis. One of them is the agreement to privilege Eurocentric ways of knowing, like quantitative fetishization, the fear of ambiguity, needing to know, and always having an answer to solve or fix something. Another is the agreement of scarcity, perfectionism, and competition. As well as pitting animals and humans against each other as opposed to being part of the same cycle.

> If you think about how many Indigenous people and minoritized people were displaced to create national parks, they have a tragic history.

Even though I still identify as a scientist because of the research I do, it's a painful title because of how extractive it is, full of all these unspoken agreements that I never agreed to. I like challenging and pushing those agreements when it comes to environmental science and the science of conservation.

When we think about conservation, it seems like national parks are the first thing that people think about. I think that national parks can be decolonized even though they have a colonial history. In the beginning, they were game parks, they were playgrounds for colonial administrators, and then they became national parks, again under the guise of conservation and environmentalism. If you think about how many Indigenous people and minoritized people were displaced to create national parks, they have a tragic history. One of the patterns that you'll find with parks is that they make a lot of money, but the people who live around them live in dire poverty; Yala National Park in Sri Lanka is a good example of this. Why not figure out benefit-sharing mechanisms in which portions of national park revenues can be shared with local residents? How do you include them in the management of national parks? There's little investment that goes into human capital development and environmental stewardship. Things like language, entrepreneurship, and tourist expectations should be taught. How can we decolonize things that are luxurious, which shouldn't be? It shouldn't be a luxury to see nature and engage with wilderness. The more access restrictions you put on those places, the more it becomes only accessible by the rich and wealthy.

WHAT WERE MY ANCESTORS DOING PRE-COLONIALISM TO CONSERVE THE LAND?

Sri Lanka herself is still experiencing a colonial hangover. We were colonized by the Dutch, Portuguese, and British, along with a penal code education system. Our environmental conservation laws are still an inheritance from British colonial times. There are still the perspectives of animals and plants over people. The idea of pristine national parks, the ideas of population control. In Sri Lanka in particular, conservation is used as a guise for racism and Islamophobia because, as I said, we are a majority Buddhist country and there's a lot of ethno-nationalism.

We had a civil war for 26 years. The way that the civil war influenced the natural landscape is a whole other thing. The environmental protection law itself in Sri Lanka is strong. About 26% of our country is protected. But what the protection means in enforcement as opposed to the written law is very different. How it's enforced in different places depends on which ethnicity and which religion lives there.

It's troubling how different foreigners and large tourism corporations

ADAM'S PEAK IN SRI LANKA
BY LAHIRU KOTTAGE

can break and amend the rules. I founded the organization Kavaya, which is a multigenerational, BIPOC group at Georgetown, exploring sustainability and self, among other things. It has me thinking about sustainability from my ancestors' perspective. What were we doing pre-colonialism to conserve and coexist with our land? Those are practices that we're still trying to unearth while we develop trust with Indigenous communities.

I think one of the biggest problems with the modern conservation perspective is that it marks a separation that I don't think exists called Terra nullius. It is the idea of land without people, and it has colonial origins.

The modern conservation perspective does not seem to realize the symbiosis and what causes human-animal conflict. Climate change has caused severe drought in Sri Lanka, causing human-elephant conflict. It intensifies things: when elephants start to look for water, they come into the cities. It's tragic because it is village dwellers, who don't have water, who will take massive tankers and try to fill in a little pond with water just so that the elephants have it.

Historically, there have been communities that have lived and coexisted with the land in ways that aren't harmful to it and help the natural cycles of the environment. We've caused climate change, and it's worsening, and now people, landscapes, and animals are experiencing the consequences of it.

The people who get to ignore that, who say, "I don't care about the people," those people have privilege and aren't part of the Global Majority experiencing severe monsoons, extreme floods, droughts, wildfires, and everything that comes with them. They also don't have a historical/ancestral connection to stewarding the land, hearing her and feeling her. To have all of that ignored is hurtful.

LEARN MORE

LubDub Theatre Company
lubdubtheatre.com

ashaneekottage.com

kavayapress.com

instagram.com/kavayacollective

Ashanee calls Sri Lanka home but came to the United States to attend college at Georgetown University. She wanted to create her own destiny and story by mixing art and science to create her unique way of communicating the sometimes difficult-to-understand concepts of science to audiences. Her work and collaborative performances with other artists are truly fascinating.
—DUDLEY EDMONDSON

A PEACOCK PHOTOGRAPHED IN SRI LANKA
BY LAHIRU KOTTAGE

DR. LORENA RIOS MENDOZA

ENVIRONMENTAL SCIENCE & CHEMISTRY PROFESSOR

SUPERIOR, WI

Dr. Lorena Rios Mendoza is an expert on microplastics. By now, we all know the environmental concerns around microplastics, including the infamous garbage patches in the oceans. After spending time in her lab, I learned about microplastics in shampoos and other widespread products, and it was unsettling, an oh-my-God moment. I've since tried to heed her message to use less plastics in favor of more sustainable options.

—DUDLEY EDMONDSON

DR. MENDOZA SMILES IN ONE OF HER RESEARCH LABS AT THE UNIVERSITY OF WISCONSIN-SUPERIOR.
BY DUDLEY EDMONDSON

MY STUDENTS CLAIM I'M A VERY HARD PROFESSOR

I teach in the chemistry and physics program at the University of Wisconsin. I have a bachelor's in chemistry and a master's in environmental science. I also have a PhD with a focus in marine electrochemistry. My students claim that I'm a very hard professor. I tell them, "If you want to work with me, then you need to work, not just work physically in the lab. You need

DR. MENDOZA HOLDING A PIECE OF OCEAN WASTE ABOARD THE LEGENDARY *ALGUITA* IN 2007; THE VESSEL HELPED UNDERFUNDED RESEARCHERS AND STUDENTS STUDY THE SUBTROPICAL GYRE IN THE NORTH PACIFIC.
COURTESY OF DR. MENDOZA

to understand what you're doing. This means you need to read, you need to study, you need to explain to me what you're doing. I don't want you to work like a robot or just follow the cook's recipe." Later, when they're going to graduate school or to work, they send me an email or call and say, "Thank you. Because you were so hard on me in class, I'm better in graduate school or on my job."

AS A KID, MOVING TO MEXICO CITY GAVE ME DIFFERENT POINTS OF VIEW

I was born in a very small town in Chihuahua, Mexico, which is close to Texas. I have six siblings. We have a very traditional Mexican family. I have happy memories from when I was a child. We moved to Mexico City when I was 9 years old, and it was a huge change for me. We moved from a small town to a big city with a population of 25 million. I learned how to survive there. It was nice to live in the big city because it gave me different points of view. I was very young, but I still learned a lot.

Studying at the biggest university in Mexico City made me what I am now. When I started studying for my master's degree, I moved to California and collected samples in the Sea of Cortez. That area is especially rich in productivity and food for marine life. Then, I had the opportunity to go to Antarctica and the North Pacific Gyre to collect samples. The North Pacific Gyre is one of many circular currents that thread through oceans across the globe. I was studying for my PhD when I was there. That's when I really started to know the ocean and wanted to start broadening my interest to study plastics. I began to focus on their negative impacts, specifically the adsorption of toxic compounds from microplastics in the environment.

WHEREVER YOU HAVE HUMANS, YOU HAVE PLASTICS

I lived in California before I moved to the Great Lakes in Superior, Wisconsin, in 2010. Some people ask me why I moved from California. I'm not crazy, but honestly, I love the cold weather.

It's been almost 20 years since I started to study plastic debris in California. It was difficult to convince people that plastics are not just garbage floating on the surface of the water; they are something else. People don't understand that plastic pollution is a bigger deal in the sediments below the surface.

When I moved to Superior, I wanted to study the plastics in the Great Lakes. I remember my

A DEAD NORTHERN GANNET, WASHED UP ON A BEACH AFTER BEING TANGLED IN A PLASTIC FISHING NET
BY ANDREW BALCOMBE/SHUTTERSTOCK

colleagues saying, "We don't have that problem here. We're not like California. We don't have contamination." I said, "Okay, just let me try." Of course, I found it, because wherever you have humans, you have plastics.

I presented those results at the America Chemical Society Conference in 2013. The reaction from the people was huge. I got a call from a New York lawyer asking me if my results were correct. Later, I had many interviews about what's happening in the Great Lakes. It was good because people responded quickly, faster than in California. What I discovered was that the number of plastic particles increased the closer I got to the wastewater treatment plants. The treatment plants only clean the water of bacteria and then send it to rivers or lakes. Here in the Great Lakes, the only explanation of where micro particles of plastics, called microbeads, were coming from was people's soaps, shampoos, or facial scrubs, among other things.

Plastics are in our cosmetic products. Before plastics became available, manufacturers used the seeds from fruits. They crushed the seeds, but they needed to dry and be kept bacteria-free. That's too many steps and is too expensive. Plastic can be used however you want, with no need to care for or keep clean, and it's a lot cheaper. Companies can even ask the plastic industry to produce the microparticles in specific sizes. We know that plastic is not biodegradable. This means there are not any bacteria that can cut the big chains of these polymers. That is the problem.

> We don't really have recycling in the United States. In the United States, recycling is just cutting it, packaging it up, and sending it away.

WE DON'T REALLY RECYCLE IN THE UNITED STATES

The other problem we have is the numbered triangles, 1 to 7, on the bottom or sides of plastic containers. This is theoretically for recycling. However, this is misleading because plastic is not easily recyclable and can stay for thousands of years before partially breaking down into small fragments. We have more than 5,000 different kinds of plastics. Those recycling symbols only apply to seven kinds of plastic. We don't really have recycling in the United States. In the United States, recycling is just cutting it, packaging it up, and sending it away.

The United States and other rich countries have started to see the problem because China and other parts of Asia will no longer accept their garbage. Now, they've started sending it to Mexico, and Mexico is saying, "Yes, we can receive your garbage, recycle it, and produce energy by burning it." The truth is that they will produce more contamination. I'm trying to talk with them and say, "Be careful; this is not the solution." I am working to train people in the sampling and analysis of microplastic particles.

THE PACIFIC GARBAGE PATCH IS THREE TIMES THE SIZE OF TEXAS

The Pacific Garbage Patch is an accumulation of plastic items like containers, buoys, bags, fishing nets, buckets, crates, and other unidentifiable fragments of plastic in the ocean. It is dispersed in the water column, and it is not very dense due to its huge size: the Pacific Gyre is around three times the state of Texas. The patch formed because

PLASTIC POLLUTION COLLECTED FROM MIDWAY ATOLL, NOT FAR FROM THE NORTH PACIFIC GYRE
NOAA/NMFS/PACIFIC ISLANDS FISHERIES SCIENCE CENTER BLOG/KRISTEN KELLY

of the physical conditions in the ocean, like high-pressure systems and different salinity and temperature. This gyre is always moving and is just one of five in the world's oceans. Any plastic that makes it there is trapped.

Unfortunately, some animals think it's food and are eating the plastic. Normally, red plastic is the first to be eaten by birds and other marine life because animals confuse it with squid. Plastic bags can get mistaken for jellyfish by sea turtles. And small plastic pellets look like eggs, and some organisms eat them.

CHEMICALS IN SINGLE-USE, DISPOSABLE PLASTIC BOTTLED WATER CONTAINERS CAN CAUSE CANCER

We use two main kinds of plastics: polyethylene and polypropylene. These plastics can float in water; this is why we can see them on the beach and on the surface. But all the rest of the plastic goes down to the sediment in the lakes and rivers or in the ocean. Some people think we can heat it up and turn it back into oil again and make new plastic. That isn't a good idea because the plastic industry puts in too many chemicals, and some of those chemicals can produce more pollution. For example, disposable plastic bottles have chemicals that can cause cancer, called phthalates. People are drinking that water, thinking it's safe, but if the plastic bottle was in the sun outside or heated up in some way, it's leaching out chemicals into the water. The heat essentially conducts an extraction, like I would in the lab. If you warm your plastic containers in the microwave, you are doing the same, and the chemicals may end up in your food.

We need to understand that humans are the main source of plastic waste. Be conscious and try to avoid the use of plastic items. Try this in small steps and use less plastic whenever you can. It may be painful, but it is possible. The more educated people are about it, the more people in power will have to take action. Remember, if we buy less plastic, the plastic industry will produce less plastic.

LEARN MORE

NOAA Sea Grant
seagrant.noaa.gov

Dr. Rios Mendoza teaches at the University of Wisconsin, Superior. Our conversations during her interview really opened my eyes about the problems with microplastics and the many sources of this pollution that many of us don't think about. It has certainly changed the way I shop and the items I no longer buy as a result. Once you are fully aware of how big the problem is, it may seem overwhelming. The bottom line is we have the power to turn this thing around, and it starts with buying less plastic.

—DUDLEY EDMONDSON

JARS OF PLASTICS IN DR. MENDOZA'S LAB, COLLECTED FROM THE NORTH PACIFIC GARBAGE PATCH
BY DUDLEY EDMONDSON

DR. SEBASTIAN ECHEVERRI

SCIENCE COMMUNICATOR

SOUTH ORANGE, NJ

Dr. Sebastian Echeverri is a spider expert and very focused on them. He's documented some amazing things about jumping spiders, his favorite group, including how they communicate. He's also dedicated himself to environmental education and working with underserved communities. As an expert who is also a young person of color, he's having a great impact.

—DUDLEY EDMONDSON

SEBASTIAN HOLDING A TARANTULA, ONE OF THE MANY ARACHNIDS IN HIS COLLECTION
BY DUDLEY EDMONDSON

MAKING SCIENCE FUN, ACCESSIBLE, AND INCLUSIVE

I have a PhD in Ecology and Evolution from the University of Pittsburgh. I'm a scientist, a science communicator, and a wildlife photographer. As a scientist, I am interested in how animals talk to each other, how they see each other, and how that has evolved over time.

My research is on spiders, with a focus on jumping spiders and tarantulas. For my PhD, I looked at jumping spiders—these tiny animals that do incredible courtship displays—and how they get their

SEBASTIAN WITH HIS GRANDMOTHER (LEFT) AND HIS MOTHER BEHIND HIM ON VACATION IN UPSTATE NEW YORK
COURTESY OF SEBASTIAN ECHEVERRI

audience's attention when they are about to show off their fanciest dance moves. Now, I am studying how tarantulas' eyes have evolved as different species have adapted to living in different habitats.

Parallel to that, I had been working on a lot of science communication throughout grad school, which is now the bulk of my career. As a science communicator, my job is to make science inclusive, accessible, and interesting in ways that are fun for people.

I got into spiders through a meeting with a professor. It was some offhand thing he said like, "Let me show you some research that's not on my website yet." Then he pulled up a video of a jumping spider's courtship dance, a high-quality, up-close video. I had never seen anything like that before. It was such a shock, that I stood up and was like, "Okay, wait, tell me more about this." I was hooked. That is the kind of experience I want to build for others with my science communication work.

AS A CHILD, I WANTED TO BE STEVE IRWIN

I was born in Colombia, like my parents, and I grew up for the most part in the United States. In 1993, when I was 2, my family moved to New York City. We moved into an apartment in Queens that my paternal grandparents lived in. My grandfather owned a Colombian bakery, and we lived with them for a little bit while we were getting settled. A lot of my early childhood memories are of him and their tiny apartment and the bakery. There were some days that I would get to sit in the bakery with him and eat Colombian baked goods all the time.

My mom was a computer programmer, so she got information technology jobs and was able to support us well. And so, we were able to have our own house in Queens later on. I've learned a lot from my mom about empathy for others and for the natural world. It's something she taught me, along with her mother, who also loves nature.

One of my earliest memories was getting hooked on watching science and nature documentaries on cable TV. That was the golden age of Discovery Channel, Science Channel, and *National Geographic,* when those shows would play all day long. I have one memory of convincing my mom to let me call in sick to school so I could stay home and watch documentaries all day. And to her credit, she did let me do this at least once.

My idol was Steve Irwin. I wanted to meet him, I wanted to be him. His whole thing was that no matter what species he was talking about, he would say, "Here's an amazing animal, look how beautiful it is. It's fascinating. Let's learn about it."

I had no idea that showing people cool animals and teaching them about it was a job. I didn't know it was called 'science communication' until much later. All the people that I saw on TV doing it were white dudes, so I didn't feel like it was something where I'd be welcome. I went into "regular" science as an attempt to get close to that, and because the path seemed simple—just keep going to school. Now here I am, a lot closer to Steve Irwin's career than I ever thought I could get.

> My idol was Steve Irwin. I wanted to meet him, I wanted to be him. Now here I am, a lot closer to Steve Irwin's career than I ever thought I could get.

BEING INCLUSIVE AND EQUITABLE IN EDUCATION DOES NOT COMPROMISE ITS QUALTY

A lot of the education system, especially anything science related, is really focused on the history and perspectives of old white men. They are who we all learn about. Their ideas and perspectives get taught as fact, or as examples in a way that erases the very real and

SEBASTIAN EDUCATING A GROUP OF YOUNGSTERS ABOUT SPIDERS AND OTHER INVERTEBRATES
COURTESY OF SEBASTIAN ECHEVERRI

important contributions of everyone else and limits the way that students can approach problems. It also limits the way that you can interact, feel, and connect with what you're learning in your field.

The RIOS (Racially-just, Inclusive, and Open STEM) Institute is an organization that I worked for as communications manager. They're a wonderful group of scientists and educators who are trying to implement changes within educational institutions. They're trying to change how we approach STEM education by focusing on the experiences and needs of people who

STEM education has previously ignored, many of whom are people of color. Their entire thing is about breaking down the various barriers that people face in higher education.

The idea that we have to somehow compromise on the quality of education to be inclusive, or to be just and equitable in how we teach, is wrong. The real world is complicated, and real biology is complicated and is best understood from a diverse set of perspectives.

In the early 1900s, the first scientist to discover hearing in insects and to revolutionize experimental design was a Black man named Charles Henry Turner. Before that, people didn't know that insects could hear airborne sounds, or at least there was no evidence for it. Turner did a rigorous series of experiments at a time where "experiments" could just mean "I think I saw something, and I wrote it down." He proved that not only could insects hear, but that they heard certain pitches more than others, and they could even learn to associate things with certain sounds. It was big. Turner was an incredibly important entomologist, but we're never told about people like him because white Eurocentric perspectives dominate the education system and educational media.

THE INTERNET IS A GATEWAY TO SCIENTISTS OF COLOR

There's a lot of things about our current internet age that are messed up. But one of the great things is that scientists, wildlife experts, and science communicators of color are on the Internet, and you can finally find them. Many of them have YouTube channels, Twitter feeds, and Instagram content. It's exciting because now the barriers to finding inspiration have gotten smaller.

AN ADHD DIAGNOSIS DID NOT STOP MY EDUCATIONAL PURSUITS

I learned that I had ADHD about a year after I had finished my PhD. So, if you ever feel that you can't do stuff because you have ADHD, consider I got the highest educational degree possible without knowing that I had it. Granted, ADHD made it really, really, hard in ways that were not visible to most people. People with ADHD find ways to cope and have to do a lot of extra work in their heads. It didn't get noticed early on because I still did okay in school. The way that you're treated as a person of color with ADHD is different than the way you're treated if you are a white person. If you look at who's being diagnosed and treated, a lot of the treatment was particularly focused on white boys. They were getting medication early on because doctors were trained to look for symptoms most unique to them.

The thing about having any sort of disability is that it does not necessarily disqualify you from achieving what you want, but it can make it harder, and I don't want to pretend that it doesn't. The reason I wanted to get a diagnosis was because I was feeling like I couldn't handle my day anymore. The symptoms and the conditions that people with disabilities have are real. But the experience of being disabled is influenced by society. For the longest time I had terrible vision, but I wouldn't be considered disabled because I could just put on glasses and be fine. We have an entire system for checking people's vision and making glasses easily available, in addition to creating ways to make the world work for them.

In the society that I hope we're building, we will have similar support systems for people with other conditions. I learned I had ADHD after I went to a psychiatric specialist for an adult autism diagnosis and they were like, "Yeah, you probably have that, too, but you also tested ridiculously high for ADHD, and you should look into that."

So, if you're autistic and have ADHD, you need different types of support than what many "normal" people in the world are already getting because society is built to support their needs and not yours. Which means you just need the equivalent of glasses, and the understanding that you deserve to be supported like every neurotypical person. We're raised in a world that makes us feel different because we need extra stuff. But it's because the world we're in is focused on the needs of people that are not like us. And everyone deserves to have their physical and mental needs supported.

LEARN MORE

BBC Earth Podcast
bbcearth.com/podcast

Crash Course Zoology
on YouTube

Spiders of the United States & Canada
Adventure Publications, 2024

Sebastian is a young, gifted Brown science communicator. He wants to make science more accessible to the public in a fun and easy-to-understand way. His extensive research work with spiders seeks to understand how they communicate with one another. He really enjoys teaching, writing, and educating people about nature. He, too, is an outstanding role model for young BIPOC interested in a career in environmental science.

—DUDLEY EDMONDSON

SEBASTIAN IN HIS HOME OFFICE, LOOKING THROUGH HIS MANY SPECIMENS OF ARACHNIDS
BY DUDLEY EDMONDSON

CHARLIE "MACK" POWELL

FOUNDER OF PEOPLE AGAINST NEIGHBORHOOD INDUSTRIAL CONTAMINATION (P.A.N.I.C.)

BIRMINGHAM, AL

Charlie "Mack" Powell is doing environmental justice work to shut down a plant that makes coke—part of the steel-making process—creating a lot of waste and pollution in an area where the majority of folks are low-income and Black. They are surrounded by industrial plants and seem to have been targeted because they don't have the political or economic power to fight back. Working through his organization, Mack is trying to get some justice for his community.

—DUDLEY EDMONDSON

CHARLIE POWELL STANDS IN A VACANT LOT IN A NEIGHBORHOOD ACROSS THE STREET
FROM A FACTORY PRODUCING INDUSTRIAL WASTE.
BY DUDLEY EDMONDSON

ONE OF MANY HOUSES IN A NEIGHBORHOOD ACROSS THE STREET FROM AN INDUSTRIAL PLANT
BY DUDLEY EDMONDSON

MY LIFE, MY WORK

I am the founder and president of P.A.N.I.C., which is People Against Neighborhood Industrial Contamination. P.A.N.I.C. is currently fighting a battle with Bluestone Foundry Coke. Bluestone Foundry Coke turns coal into coke. They sell the coke to foundries and smelt iron with it. Coke is one of the hottest forms of fuel you can get. There's been a plant in this location for over 100 years. Right now, the plant is closed. When it's not closed, it's idling. I've been into this for 10 years now, fighting for justice. I run a nonprofit organization. I got a very low budget. Every now and then I get donations, but I put my own money in this here because this is what I believe in. What we really want is relocation buyout and compensation for the ones who want to stay. What we want is justice. The city is suing the plant for operating three years without a license permit. I think the health department is suing them too.

I GREW UP EATING CONTAMINATED FOOD BUT DIDN'T KNOW IT

I grew up in what they called Riggins. Now it's called Fairmont. Everything was fun around there when I was coming up. We didn't know any better. We had springs and wells but no running water. The industrial plant was there then, but back then it was called U.S. Pipe. During my childhood, there were plum trees, apple trees . . . everybody had a garden. But late in the evening, it would be a film coming out the sky, coming from the pipe shop. It smelled like rotten eggs. And at the same time now, all this was going in the ground, contaminating the soil, and we didn't know. Now, they had pasture around there they called the dairy farm. The cows grazed in the pasture. They even sold us the milk. We drank the milk. So now we're all contaminated, but we did not know. In the summertime, I always thought it was impossible for us to starve. There was too much running around here. My folks had a little farm. If I'd be walking down the street and you got a garden in there and you got tomatoes, I'd go in there and get me a couple tomatoes and walk on down the street and keep eating them. They didn't mind. They just didn't want you tearing it up. But we was eating all this contaminated stuff and didn't know it.

FIGHTING FOR JUSTICE, THE PEOPLE'S ASK

The city came up with what they call the "Big Ask," and that Big Ask is them coming up with $37 million. They're talking about what all

they are going to do and what they ain't going to do when they get to that $37 million. They're leaving about $12 or $19 million to move three communities—that's not enough money. We have approximately 2,500 affected people. Not only that, you've got churches involved. Some of those churches have big mortgages—$37 million ain't going to move them. The way I trickle all that down and do my math on it, that means for those houses, they're paying maybe $10,000 per house. I don't think that's enough money to move, even if you had 2,500 dog houses. To me, I think what they are asking the people to do is give up their homes for next to nothing. Our strategy at P.A.N.I.C. is to come up with the "People's Ask." We're going around in the community and asking them, "Do you want to move? Do you want to stay? Do you want to be compensated? How much you think your house is worth?" We are not trying to sell the people's houses. We want the people to sell they own houses. We're canvasing the neighborhood, asking the people those questions, then we're taking it back to the city. When the city finishes this up, the people wind up giving away a $60,000 house for $10,000. Man, that is ridiculous.

Now, some houses are worth more than that. What they did there is people over-fixed their houses. The reason that happened was because Black people had a limited amount of places where they could stay. So, a man's home was his castle, and they fixed it up like that, not knowing they will never be able to get what they put into it out of it.

And this is a shame for a city as big as Birmingham. I can't see why they will allow this to go on, now that we are able to live anywhere we want to. This is the new generation now. The generation behind us didn't know what we know now, and the generation that comes after me is gonna know more than I know, but hopefully we can get this thing settled.

This thing was supposed to have been on the Superfund site in 2012, and to be on the Superfund site, you had to score at least 25. We scored a 50.

A STRUGGLING COMMUNITY WHERE YOU DON'T KNOW WHO TO TRUST

They done lied to the people so much, the people don't know what to believe. They don't want to trust nobody and probably don't even trust me, but they got a little faith in me, more than they got in the folks coming in here. When you got the mayor, the city council, all of them leading the people wrong, what is it for them to think?

Wouldn't it be more economical if they was to move the people out and let that there be an industrial zone? You got 400 acres of contamination there and the plant ain't even started back up. Now start thinking about what happens once they stop idling and get back in full force.

Now, when they first put that thing there, the Black people who lived in the area had good jobs, good opportunities, but deep down within, they was hurting the people. I want to believe that they didn't know this, but now they know. Why don't they straighten this injustice out? I understood that the thing was supposed to have been on the Superfund site in 2012, and to be on the Superfund site, you had to score at least 25. I think we scored a 50. They said we should have been out of here. So, what is the problem now? Over the years, what these companies used to do when an activist like me is coming in, they'll go bankrupt and sell. I think what happened to the site owner was they was supposed to come in there and run it into the ground and then leave, but they got caught cause we was there. P.A.N.I.C. was there. We started putting up air monitors to see was things contaminated cause a lot of this stuff they were saying was not contaminated

MR. POWELL STANDS IN FRONT OF P.A.N.I.C. HEADQUARTERS, WHERE HE WORKS WITH VOLUNTEERS IN THEIR FIGHT AGAINST INDUSTRIAL CONTAMINATION.
BY DUDLEY EDMONDSON

was. They said that our church wasn't contaminated. Well, we put up monitors ourselves and found that it was. They came up there and cleaned up one side of the church, but how can one side of a church be contaminated and the other side ain't? Then think, you had a plant there for a hundred years doing this in different names and they dig down about 12 inches for contamination. You got a hundred years of stuff going on like this here, and them houses was built in the 1940s. Don't you think they should have went down more than 12 inches?

Oh man, come on, we ain't just fell off the cabbage truck.

LIVER CANCER, COLON CANCER, STOMACH CANCER: ALL MY FRIENDS, NOW THEY GONE

In my living room, I have a bookshelf, but I never intended to put books up there. Those pictures are people who died of cancer, strokes, and heart attacks throughout the years, but I can't get no more of them up there, and I got about 200 more in my desk drawer. My wife now has cancer. She had colon cancer, and now she has liver cancer.

I just had a nephew we buried, maybe about four months ago. He had stomach cancer. My nephew before him, he had liver cancer. One of my friends next door to him, he just had stomach cancer. I can go on and on with this cancer stuff, and there is no way you can show me that there ain't no kind of connection. Been knowing it for years and ain't thought no more about it till I got to be an activist and started looking into this here. It was about 16 of us, and now only 6 left. Yeah, all of my friends, now they gone.

> It was about 16 of us, and now only 6 left. Yeah, all of my friends, now they gone.

BLACK OR WHITE, EVERYBODY IS HUMAN

You would never see a coke plant in a predominately white neighborhood. Now we do have white people here. I hung out with a lot of the white guys. I mean, I have friends, all nationalities, but they don't come to the meetings. To me, they kind of feel like we felt in the '50s. They stay off to themselves, but they won't come to a big gathering. They want to be left alone, and it's sad cause some of them are living worse off than we are, but they still my friends. I wouldn't care if you stay in a chicken coop. That's your house, and it should be worth more than a chicken coop. I never forgot my roots, where I came from, you know. I can be poor and Black and hang out with poor whites. I look down on nobody. Everybody is human. I've been around for a long time, man, and I'm going to be around until the duration, until we get this matter resolved.

LEARN MORE

P.A.N.I.C.
facebook.com/PANIC.Bham

Charlie "Mack" Powell is a tireless fighter for justice in his community. He is the founder of P.A.N.I.C.—People Against Neighborhood Industrial Contamination. He works off a shoestring budget using his own money and small donations. He often works with university students and volunteers to do the work to help his community fight for environmental justice.
—DUDLEY EDMONDSON

ONE OF THE FACTORIES THAT P.A.N.I.C. IS UP AGAINST IN ITS FIGHT AGAINST INDUSTRIAL POLLUTION
BY DUDLEY EDMONDSON

REVEREND EDWARD PINKNEY

ENVIRONMENTAL JUSTICE ACTIVIST

BENTON HARBOR, MI

Reverend Edward Pinkney lives in Benton Harbor, Michigan, a city with lead pipe problems in the community. Edward has helped distribute clean bottled water, get the lead pipes replaced, and come to his community's aid to help them get justice and clean, safe water.

—DUDLEY EDMONDSON

WORKING FOR FAIRNESS

I'm the president of the Benton Harbor Community Water Council, a group of concerned citizens who believe clean drinking water is a human right. I make sure that I'm on top of everything that needs to be done. We've got a strong team. We got teachers, doctors, we've got everybody on our board. I'm also the president of the Black Autonomy Network Organization. The number one thing we do is court-watch. I specifically talk to every client that goes to our local courthouse to make sure that they understand the process. We know that the justice system is not always fair to us, so we help them and make sure that everybody at least gets a fair shake.

GROWING UP IN CHICAGO

I grew up in Chicago with seven brothers and sisters. My dad was a chemist. My mom was a stay-at-home mom. We lived a pretty normal life. The Fillmore police station was right behind our home, and when someone in our community got in trouble my dad would go and get people out of jail, with no money. In those days, you didn't have to bond people out of jail; you could have someone released by your name. He would put on his clothes and meet them at the police station. I felt that was one of the most remarkable things that ever occurred during my childhood. It kind of framed my life.

Baseball was my thing. I went to school on a baseball scholarship. My brother was a pitcher and my dad would help us practice. He would show us how to hold a ball and bat. He also made sure that we were good students. When he came home from work, we would have to have our homework completed and on the table because he wouldn't eat his dinner until

REVEREND PINKNEY'S FATHER WAS A CHEMIST AND AN UPSTANDING COMMUNITY MEMBER.
COURTESY OF REVEREND PINKNEY

REVEREND PINKNEY TURNS ON THE TAP IN THE COMMUNITY CENTER AND DISPENSES A GLASS OF LEAD-CONTAMINATED WATER.
BY DUDLEY EDMONDSON

he read our homework. I thought that was great. When I became a father, I started doing that with my children.

WATER PROBLEMS FOR THE CITY OF BENTON HARBOR

One day a friend called me concerned about the color of her tap water. Her daughter was in town and after doing some housework, she wanted to take a bath. When she ran the bath water, it had particles in it. The water was gold colored. She called me, and I said, take two quarts of water down to city hall and have it tested for lead. She took it to city hall, gave it to the mayor's office, and asked to have it tested. I recall that after a month passed, we decided to form the Benton Harbor Community Water Council. Our job was to find out why our water was contaminated

THE BENTON HARBOR WATER TREATMENT PLANT
BY DUDLEY EDMONDSON

with lead, but the community was never notified. I took two more quarts of water from my friend's home and sent it to the University of Michigan's biological lab. They tested the water and discovered it had very high lead levels. When we formed the water council, we challenged the city council because we weren't getting action from the city or the State Department of Environment. They're a department of the state government. The reason they said they didn't report it was because no one ever completed any water sampling. You're supposed to have 60 samples. They got partway there in 2018, when they got multiple samples. That was done by the Michigan Department of Health. If you have over 15 parts per billion, that's the danger level. They found lead levels as high as 490 parts per billion but were slow to tell the community. The Community Water Council went to work, calling people up, making sure they collected water samples from their faucets. We told them how to prepare the samples. We would drop off the kit at the front door and tell them to leave it back out on the porch so we can pick it up. We did that for three years. When lead levels were too high, the state-run Benton Harbor water filtration plant manager asked the water council to go out and get samples in hopes of finding lower levels to make it appear there was no water issue. We didn't know that was their plan, but when we found out we had to blow the whistle on them. That's when the water council filed a petition with the EPA.

This is a shame more than anything else: they were being dishonest because the city of Benton Harbor is over 90% African American. If this was a white community telling the world that these lead pipes were killing their community, state and federal officials would rush in and make a commitment to fix it. But by being a Black community, and not just this Black community, but every single Black community in America that has lead pipes got lead in the water. That's a fact. You don't have to be a Rhodes Scholar to figure that out.

THE TALE OF TWO CITIES: ONE BLACK, ONE WHITE

Saint Joseph, Michigan, just across the St. Joseph River, has safe drinking water for their community that's over 90% Caucasian. They broke off and got their own water filtration plant in 2019. Other surrounding suburban townships that used to get water from the Benton Harbor's water filtration plant did the same. That had an economic impact on the city of Benton Harbor because we lost those customers and that revenue. I don't know if it was designed to do that, but in many ways, it helped to destroy the water plant itself.

CLEAN WATER IS A HUMAN RIGHT

Once the problem was uncovered, we started getting water donations from just about everybody. We had lots of people donating money through our GoFundMe fund. So many different organizations were funding us. After we filed that petition, we started getting free water from the state. The state also gives us, the water council, $1,500 dollars a day to pay teams of people to distribute water in our community. We put cases of water on folk's front porch or steps. We'll also take it in the house for our community elders because we have teams who know the families. We try to do the whole city. We

We try to do the whole city. We try to give everybody enough water to last a week. If folks need more, we'll deliver that too.

try to give everybody enough water to last a week. If folks need more, we'll deliver that too.

> Basically, they were saying if you are Black, and you got lead in your water, then it's okay, that's the message I got.

ENVIRONMENTAL JUSTICE FOR THE PEOPLE OF BENTON HARBOR

The pipes are being replaced, which is what we were striving for. If it weren't an election year, I am not sure things would be happening, because they just don't move that fast for Black people. My biggest concern is, is it being done correctly? I do have members of the water council on the ground checking. Also, we are working with them to repair the water filtration plant. It took us four years to get to this point. Four years of fighting and trying to convince city, state, and federal officials that the water was contaminated with lead and bacteria. They couldn't admit that there was a problem, even though they knew it, because of the high cost. Basically, they were saying if you are Black, and you got lead in your water, then it's okay, that's the message I got. The only thing we want is safe, clean water for this community. Everything we do is for our children. That's our future, and that's where I want to lay the foundation.

LEARN MORE

Reverend Edward Pinkney
Benton Harbor Community Water Council
bhcwc2.org

Black Autonomy Network Community Organization
bhbanco.org

Reverend Edward Pinkney is a determined, mission-driven man. He is determined to ensure that the people of Benton Harbor, Michigan, have safe, clean drinking water. He looks out for his community first, whether it's community members navigating the court system or efforts to address the state's ongoing drinking water problems. He is up for the fight—for justice and the respect that everyone in his community deserves.

—DUDLEY EDMONDSON

A TEAM OF COMMUNITY MEMBERS DELIVERS WATER TO THE RESIDENTS OF BENTON HARBOR, MICHIGAN.
BY DUDLEY EDMONDSON

AS THE PLIGHT OF FLINT AND BENTON HARBOR SHOW, LEAD-TAINTED WATER IS A WIDESPREAD PROBLEM.
BY IXEPOP/SHUTTERSTOCK

WHAT?!

"Chronic lead exposure in public drinking water remains a persistent environmental injustice that disproportionately affects Black communities across the United States."

—NIGRA, ET AL, *JOURNAL OF THE AMERICAN SOCIETY OF NEPHROLOGY*, 2021

SHARON LAVIGNE

ENVIRONMENTAL JUSTICE ACTIVIST AND FOUNDER OF RISE ST. JAMES

ST. JAMES, LA

Sharon Lavigne has been fighting industrial waste and industrial pollution in Cancer Alley in Louisiana. Her community is one where industry and government have treated the community as a sacrifice zone, all in the name of capitalism. A devout Catholic, she is called to this work to empower people and help save her community, friends, and family.

—DUDLEY EDMONDSON

SHARON STANDS IN HER FRONT YARD WITH HER GRANDSON BETWEEN SIGNS
PROTESTING POLLUTION IN HER COMMUNITY.
BY DUDLEY EDMONDSON

A VOICE FOR THE PEOPLE OF CANCER ALLEY

I am Sharon Lavigne. I live in Saint James Parish. I was a special education schoolteacher here. I taught school for 38 years. I lived in a little area called Chatman Town. That's where I was born and raised and have lived here most of my life until 1977. Now, I am living 5 miles up the road in Brooks Town.

I am the founder of a community-based Environmental Justice organization called RISE St. James. The mission of RISE Saint James is to save the lives of the people in Cancer Alley, but especially Saint James Parish. And to stop any industry that's trying to come in here to harm our people and to stop the expansions of the ones that's already here.

I have a whole list of chemicals these industries are emitting into the air and water that's cancer-causing or harmful: ethylene oxide, formaldehyde, sulfur dioxide. Industry expansion would be a death sentence for Saint James Parish.

AS A CHILD, I REMEMBER MY DADDY WAS A BRAVE MAN

When I was a little girl, we had gardens of vegetables. We lived off the land. My daddy had cows, hogs, chickens—we even had ducks. My daddy would milk the cows in the morning and put the pot on the stove and boil the milk. The top layer would have a cream. He would take the cream off and get enough cream and make butter.

My dad was a sugar cane farmer. That was his livelihood. He also was a school bus driver. After he helped integrate the school in 1966, Saint James Parish, Sugar Co-op stopped receiving his sugar cane because of the civil rights work that he did. My daddy was brave, and he fought for civil rights in St. James. He was the President of the NAACP chapter in the Fifth District of St. James.

RACIALLY INTEGRATING THE SCHOOLS OF SAINT JAMES

When he integrated the school, the superintendent wouldn't allow us to go because we were too old. He wanted the young children around 6 years old. The little babies. They felt that the bigger ones might fight back or something when people called you the N-word. The little ones don't know about the N-word, that kind of stuff.

So my daddy hustled up seven children. He went to the homes, door-to-door, and asked if they had a child that was old enough, to see if they would allow their child to go. Some of the parents said no; seven

SHARON'S FATHER, MILTON CAYETTE, SR., STANDING WITH THE SEVEN MOTHERS WHO ALLOWED
MILTON TO INTEGRATE ST. JAMES HIGH SCHOOL WITH THEIR CHILDREN

COURTESY OF SHARON LAVIGNE

said yes. When they went to the school that day, the mothers came with him. He couldn't get the men to come with him. The men had all kind of excuses. The janitor at that school said he was so afraid for my daddy when he walked off that bus with a cowboy hat on his head. He said he was so afraid somebody would shoot him, so he prayed for my daddy when he walked off the bus with seven children and their mothers.

After that happened, I guess things changed. The white people was against my daddy after that, but he didn't care. Then when all this happened with Dr. King, my daddy received threatening phone calls.

I remember him talking to my mama, asking her, "What's going on?" I remember that they threatened to burn our house down. My daddy was Catholic, just like I am. He walked around the house with his holy water, sprinkled holy water, and he prayed. I think I know what he probably said, Dear Lord, you gave me this home. Protect my home and my family.

They didn't burn the house—they burned his truck. His truck was parked in Donaldsonville, where he would pick the kids up in a school bus and take them to school. Then he would come back home in the daytime in the truck. He received a phone call saying, your truck is on fire. He didn't go around the truck with the holy water. He went around the house. The house was spared.

I DIDN'T REALIZE WE WERE A SACRIFICE ZONE

When industry first came in here back in 1967, or '68, somewhere around there, they said that they were going to be good neighbors. They brought jobs. Everybody was excited. Little St. James is on the map, boy, we got a plant here. I used to hear folks say that, not knowing the other side of that plant, that it brings pollution too. They didn't tell us that. Then they asked to expand.

All these industries coming over here, but not knowing we were a sacrifice zone, we sacrificed our lives so they can make money.

The parish said, "Okay." Then another one came, then another one came. All these industries coming over here, but not knowing we were a sacrifice zone, we sacrificed our lives so they can make money, and our public officials are allowing it because they are getting something too.

I found out that one and maybe more of the companies around us, just a few miles from my home, have been polluting the air and water with toxic chemicals. We've been breathing and drinking those chemicals for years, and we didn't know it. So the fish are eating the chemicals. My friend Robert, who lived two doors down, his wife died of cancer, but he would go behind the levy and fish. I said, "Robert, I don't want you bringing me fish." I said, "Robert, I don't want that fish," and Robert would say, "Sharon, just cut the belly out, that jelly part, that part is not good." I said, "The whole fish is not good because of the chemicals in the water." Robert said, "Okay then, Sharon," and he would eat the fish. He wound up with throat cancer, and he died a few years ago. That was my good friend. He was eating the fish from the river.

I PRAYED FROM DEEP IN MY GUTS FOR GOD TO STOP FORMOSA PLASTICS

In April 2018, our governor, John Bel Edwards, announced a $9.4-billion deal with Formosa Plastics to build a plant here in St. James, just 2 miles from my home. I was on the phone this morning with some ladies, and we were talking and I started crying. I asked God not to let me cry because I get so emotional when I talk about it and how I got into this fight. I had no idea I was getting into all of this. All I wanted to do was to stop industry from coming into Saint James Parish.

SHARON PROTESTING IN WASHINGTON, D.C., AND HOLDING AN ENVELOPE FULL OF SIGNATURES FOR PRESIDENT BIDEN
COURTSEY OF SHARON LAVIGNE

THE LARGE SIGN THAT SITS IN SHARON'S FRONT YARD PROTESTING FORMOSA PLASTICS
BY DUDLEY EDMONDSON

All these years I was riding past these places, going to work and seeing all these storage tanks. We have over 100 storage tanks within a 10-mile radius in my community. It didn't dawn on me what all this was about. I didn't know we were being destroyed until 2015. People dying, people moving out. The white people, some of them got bought out or had buffer zones protecting them. Black folks didn't get those same protections and even were restricted from selling their own land. But this is where I want to stay. I live on family land. We have family property on the East Bank and West Bank of the Mississippi.

It's just so touching to me to know that God chose me out of all the people in Saint James Parish. I know other people were praying, but maybe I prayed with such passion. Maybe I prayed from my guts to ask him to stop Formosa Plastics

from being built 2 miles from my home and stop any of the industries from coming in and destroying our lives.

THEY KNOWINGLY PUT THESE INDUSTRIES IN BLACK COMMUNITIES

I think race is a factor. I think they choose low-income impoverished areas, and people who are living in poverty and people who are already being polluted. They think we are an easy target. Some of us are illiterate and they figure that when they come to these communities and talk to the people, they figure that they're talking over our heads and folks don't understand what's going on. These companies come in with a nice, beautiful picture and make it sound so good, and people fall for it.

They know where to put these industries because they know it's a lot of Black people. It's 95% Black in my district, and they know that people won't want to speak up. We don't have the education that we need to deal with these people and the language that they use to trick us. So it's racism. They will not go into a predominately white district, talk their talk, and get them to sell out.

The latest industry that's trying to come into Saint James is DG Fuels. They make fuel for jet airplanes. They want to build this plant in Saint James, about 5 miles from me. The CEO of DG Fuels met with the parish council members, and they all gave them the green light. Then I understand the lawyer told them you got to speak to RISE Saint James. We met in my garage. I fixed it up nice with tables and chairs. We told him, "We don't want you here." We gonna show him, they're not coming here to pollute us.

THE PEOPLE OF SAINT JAMES NEED THE SUPPORT OF A LAW FIRM

We need attorneys to help us. We need legal help. We need legal representation for the people. I want it for the people whose loved ones have died because of industry. We need a law firm that wants to help us. I want someone to do research on all of these plants in this area. We wanted to do a study. We don't have a university that would take it up.

Maybe Stanford, I don't know if they would take it up. The way I understand things, the state registry does not have records about all the cancer. By drawing blood or taking urine samples, we can prove that we are right, and the companies are wrong. We can't get the funding. Nobody's going to give us funding for that. So there's another problem we're facing.

> We need attorneys to help us. We need legal help. We need legal representation for the people. I want it for the people whose loved ones have died because of industry.

WE CAN'T TAKE IT ANYMORE, DEAR GOD; THE COMMUNITY UNITES TO FIGHT INDUSTRY

In 2016, I found out I had autoimmune hepatitis, which is dealing with my liver. I thought, If I lose my liver, I lose my life. I said, dear Lord, I can't lose my liver. I did my research and I found out that autoimmune hepatitis can come from industrial pollutants. That opened my eyes a little bit.

In 2018, myself and a group of Fifth District residents decided we need to try to fight Formosa Plastics. Of course, we were told once again that the governor approved it, so it's a done deal. People told us, "You don't fight industry, you're wasting your time." I would go to the meetings, come home and take my bath to get ready to go to work the next day, and then lay in bed, and sometimes I would cry myself to sleep. I was told we might have to move—that really stayed on my mind. I would pray to God, I was talking to him, telling him, "No more, we can't take it anymore, dear God."

I went to the meetings once a month, faithfully: me, Geraldine, and Beverly. We were there all the time. Then we asked the community, let's have a march. We had a march on September 8, 2018. We marched down a street called Burton Lane. That's where people are sandwiched in by two industries on either side. We talked to the people in Covent, Louisiana, and Saint John the Baptist Parish. They came and they marched with us. We went to this pavilion at Welcome Park. That march was my very first time speaking in public. I had my little speech written on a piece of paper. That day was so wonderful.

LOSING A FRIEND AND FIGHTER HURT SO BAD

We asked for assistance in our fight against Formosa from another local community organization we'd partnered with in the past, and they told us no. That was heartbreaking to me, Geraldine, and Beverly. Geraldine kept saying, "I'm tired of living like this, breathing this dirty air. All the chemicals in the air, it comes in my house." She wanted to get out because it was too much pollution. When the train would go by, it would shake her house, shift her front door, and she tried to get her door fixed. It was a big mess. She asked a company to buy her out so she can move to go someplace else. Geraldine heard they were buying the white people out, so she went to the company and asked them to buy her out. She got her bags packed, thinking that she would be able to get out. Her bags were packed. They told her no. She died in 2019.

SHARON SITS WITH HER DEAR FRIEND GERALDINE, WHO PASSED AWAY IN 2019, STILL HOPING TO BE RELOCATED.
COURTESY OF SHARON LAVIGNE

That hurts so bad. I went to the hospital to see Geraldine and held her hand—it was ice cold. I knew she was gone. I said, dear Lord, let your will be done dear God, I know you're calling her. Myself and all the other ladies held hands and we prayed around her bed. It was so hurtful to see our girl go, but she was a fighter.

STOP THE AMMONIA PLANT NEXT TO THE SCHOOL

Now we have to continue fighting with Formosa Plastics. They're not coming in here because of God, not because of me, not because of RISE, because God is the one that is going to stop these people. We had people from New Orleans tell us about the fight that they had in Wallace, Louisiana, when Formosa tried to build over there in the 1990s. They didn't build there and now 30, 40 years later they want to come to Saint James Parish. They think we will just roll over and let them come? I don't know who told them the people in Saint James don't speak up. If I hadn't spoken up, that plant would have been built. So God chose me to bring the word out to the people.

We're spreading out right now. We have to go to Donaldsonville, Louisiana. That's where God is leading us. We want to stop the emissions from the ammonia plant that's right next to a school. Those children are breathing it and have asthma. So many people in Donaldsonville are calling us, saying the families are dying of cancer. So that's God's calling to us and we have to go and help our people.

LEARN MORE

RISE St. James
risestjames.org

Sharon is a strong woman of faith and determination. All she wanted was to live on her family land in a community free of industrial pollution. She did not ask for this fight, but when it became obvious to her that the oil, gas, and chemical industry was determined to destroy the lives of the people of St. James, she'd had enough. She prayed to God to give her strength to speak for the community and enlist others to help in the fight. Sharon is a genuine hero in my eyes. —DUDLEY EDMONDSON

A.G. SAÑO

ENVIRONMENTAL ACTIVIST AND MURALIST

MANILA, PHILIPPINES

A.G. Saño is a person with great empathy and a true activist, and I admire the work that he's done with marine life and dolphin conservation. As of late, he's focused on climate change and how the industrial polluters and oil refineries are largely responsible, and how folks living on remote islands are suffering as a result. He spoke about his personal experience with Typhoon Haiyan, a devastating storm, but he somehow retains great empathy.

—DUDLEY EDMONDSON

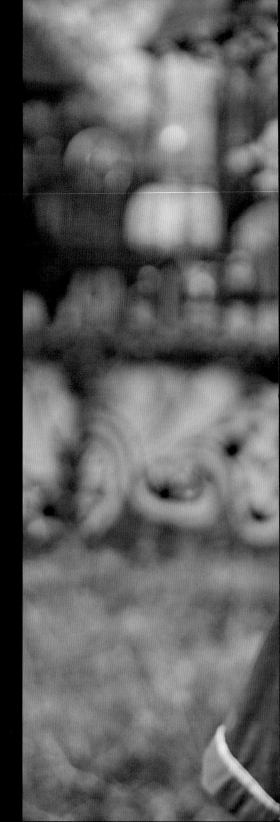

BELOVED WARRIOR

People call me A.G., but the real meaning of that is Amado Guerrero. It's a Spanish phrase that means beloved warrior. I was named after a Filipino revolutionary who fought the dictatorial government during the 1970s and 1980s. Both of my parents were victims of martial law. When I say victims, I mean they were imprisoned and tortured during the time of President Marcos. That's why they chose the name Amado Guerrero for me. But growing up, I never really used Amado because it sounded subversive. Marcos was still in power until I was 10 years old.

After graduating with my bachelor's degree in 1997, I immediately joined my professor's architectural firm. He is very prominent in the Philippines and was designated a national artist, which only a few professionals in the Philippines ever achieve. Architecture taught me how to work with nature rather than to work over nature. Around the same time, I volunteered to join WWF (World Wildlife Fund) as a photographer for humpback whale research. I haven't stopped doing conservation work since 1999.

I believe whatever I am now is because of how my parents raised me. My father made it a point to take me and my brothers on never-ending weekend escapades out into nature. He would drive us to the closest province south of Manila, where we could still see pristine forest landscapes and lakes. When we didn't have much time to go out of the city, he would take us to the parks at the University of the Philippines campus, where there is still amazing forest cover. Those trips were nearly every single weekend. Even if I hadn't done my homework, he would insist we go because it was the family tradition. I couldn't appreciate it then, but I credit those moments with my family as why I became a nature lover and an environmental defender.

MY STREET ART CAMPAIGN TO SAVE MARINE WILDLIFE

I started my street art campaign in 2010. I would mainly paint dolphins and whales because I saw the slaughters and dolphin trafficking issues in a documentary called *The Cove*. In the beginning there were only 10 volunteers to help me paint a wall in the northern Philippines where we were doing humpback whale research. After that we went back to Manila. I still felt grief from watching the documentary, so I posted on Facebook that we would paint one dolphin for

A.G. (FAR RIGHT) AND HIS FRIEND AGIT PAINTING A MURAL
COURTESY OF A.G. SAÑO

every dolphin captured anywhere in the world. I tried to connect with dolphin advocates from different regions to monitor how many are captured. Some are sold to dolphin facilities for dolphin shows. Many of them are slaughtered, and a lot more are caught by tuna vessels and other types of fishing processes. After posting that, a lot of people on Facebook showed their support, but there wasn't anyone who would volunteer their walls for me to paint. So, I was in a bit of a dilemma, but one of my close friends who is also an environmentalist with Greenpeace said, "Hey, I have a small restaurant down the road, you can paint my wall to start your campaign." We painted

the wall, and I took a time-lapse video of the process and I sent it to CNN, not knowing what would really happen. I got an email from CNN International in less than a week. One of the producers told me that she wanted to do a story about what I was doing. She scheduled the Skype interview for 3 a.m. the next day. I put my phone beside my pillow and set an alarm. When I woke up, it was bright and sunny and way past 3 a.m. When I looked at my phone, it was silent for some reason. There were at least five missed calls from a US number. I immediately emailed the producer, and she sent me questions and she told me to just record my answers.

A MURAL OF A.G.'S FRIEND AGIT, WHICH A.G. PAINTED
COURTESY OF A.G. SAÑO

The story came out and aired for one whole week on CNN. They replayed it five times a day. That's when everything exploded. People from all over the world were thanking me or supporting me. Now, 13 years later, I've worked with more than 250,000 volunteers from more than 65 countries. That's how I got noticed by communities all over the world. I have been invited to do mural sessions in Europe, the US, and Asia. A lot of big names, even celebrity personalities, have joined us to paint and show their support. What I value most is having a chance to paint with the men and the women from the streets. When people see us, there's a great opportunity to give them a brush, teach them how to paint, and make them a part of the advocacy.

CLIMATE ISSUES ARE DEFINITELY A HUMAN RIGHTS ISSUE

Dakila is a Filipino, Tagalog word that refers to heroism, to being there for others. Dakila, a Philippine Collective for Modern Heroism, was founded by different artists from the Philippines. Most of us were based in Manila, where the organization was founded. I was invited to be part of it by some of my good musician friends. We have a strong youth membership, and we are focused on human rights campaigns. We use our different forms of art to advance the campaign for human rights. Climate issues are definitely a human rights issue.

We were able to amplify the truth that climate disturbances were the result of human activity, especially from big fossil fuel companies. And that there was correlation to people losing lives, livelihood, and shelter. Those are all basic human rights. When there is a climate disaster, no one is being held accountable. Basic human rights are trampled because of the changing climate and the activities of corporations and governments around the world.

No one is held accountable. Rich companies and countries become richer by burning fossil fuels and destroying the environment, while we, the poorer countries who are not responsible, are suffering. That's the big gap that we want to address through climate action.

SUPER TYPHOON HAIYAN, THE DEADLIEST STORM IN RECORDED HISTORY

The typhoon is something that's really hard to talk about, but I choose to do so because it's a form of witnessing. I feel that witnessing for my countrymen would be something that would help a lot in advancing our cause and bringing their voices to the world. When the typhoon struck, I was visiting my father's hometown of Tacloban, far from Manilla.

I was there visiting a friend. I think it was around 2003 when I met a musician in Manila who I discovered was born and raised in Tacloban city. We became good friends. Around 2012, he decided to go back home with his wife and kid. I was happy about that because I knew that I would be able to visit him and see some of my relatives. I was able to visit him and also ask him to mentor me in the practice of traditional tattoo, which is like hand poke tattoo. He's one of the few Filipinos who does it.

When there is a climate disaster, no one is being held accountable.

Back in November 2013, Typhoon Haiyan was already brewing. It was in the news as a tropical depression in the Pacific when I booked my flight to Tacloban City. As it came closer to the Philippines, it got bigger and stronger until the weather agencies were predicting that it would be the strongest storm in human history. I remember I was on my way to the airport in Manila, and my brother, who was part of the Climate Change Commission in the Philippines, messaged me and said, "Hey, you better be careful because it's going to be really strong and it's going to hit Tacloban City." I said, "Yeah, I'll be ready." I'd been doing a lot of firefighter and rescue training, so I knew I was ready.

I had gone on that trip to meet up with my artist friend. His name's Jonas, but we call him Agit. He couldn't be with me during the day because he was doing an intricate tattoo, but he asked one of his childhood friends to take care of me. So we went to the beach, spent the day near the sea, and then during the night, we went to Agit's studio. I started documenting his work because it's amazing. After that, we went out for some drinks. I remember Agit didn't drink, but he was there. Before we said goodbye, I told him to be extra cautious because the storm was going to be big.

I THOUGHT, THIS IS HOW PEOPLE FEEL BEFORE THEY DIE

My original plan was to sleep in his house while the storm passed. I realized I shouldn't burden them with a visitor while an intense event was happening. I wanted them to focus on family and their safety. I was able to find a building in the downtown area and I stayed on the fourth floor. I thought that I was safe because, if there was a storm surge or big flood, I would be safe upstairs.

I woke up around 5 a.m. The storm was already battering the city and there were all of these weird loud noises. I realized that the roof of the building was being torn apart. That's what the noise was. I went to the window, and it was shaking. I tried to peek out, and all I could see was the roofs of buildings flying 50 feet in the air. I went to the fire escape, but the aluminum door was shaking, so I thought I should stay away. Then, just as I walked away from the door, it imploded because of the strong winds. It was torn apart, and a lot of debris started coming in. Since the roof was gone, water was coming in and flowing down the steps like a waterfall. I went back to my room for shelter, but I couldn't close the door because the wind was so strong. I pushed my bed over to the door, and then I pushed from the other wall so that I could shut the door. I was alone, and the water was coming in the bottom of the door. I got some sheets to cover it and took some deep breaths knowing that I needed to be calm, and not to panic. I made sure I had two bags, and I separated the valuables from the nonvaluables. When I got to the ground floor, which is basically a

VEHICLES STACKED ON ONE ANOTHER IN TACLOBAN CITY AFTER TYPHOON HAIYAN
BY A.G. SAÑO

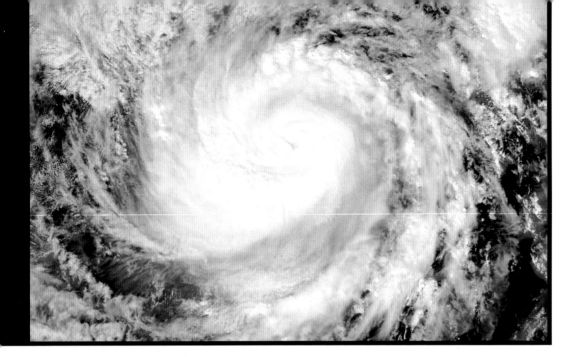

TYPHOON HAIYAN AFTER PASSING THROUGH THE PHILIPPINES
COURTESY OF NASA GODDARD

corridor to a glass door, it was filled with people who were trapped and didn't make it out. Some guy from the building asked me, "Hey sir, where do you think you're going?" I said, "I need to go to the capital because I need to communicate to my family." He said, "No, you can't, you can't go out."

I was about to go back upstairs and took maybe three steps and the glass door imploded, sending shrapnel all over. People started to shout and began running upstairs. There were women crying, and some of the building staff ushered us to the mezzanine on the second floor. There was a hall and a small room there where they could shelter us. The men started pushing shelves toward the windows so that if the windows shattered, we would be protected from the glass. The building started to shake like there was an earthquake. I thought if the projections were right, then the waves could be up to 30 feet high. I went up to the third floor to the corridor to check if it was a good place to hide. I checked the fire escape door at the end of the corridor, which was still intact. There was a couch in the corridor where there was a man who was in shock just sitting there. I was talking to him, but he wasn't listening. I sat down beside him and tried to comfort him. The building started to shake again, really violently. This happened on and off for 2 hours.

The people from the mezzanine came up to that same corridor and it filled with people.

At some point I started praying, but my prayer was not to be saved because I had already resigned myself to my fate, that maybe this is it. I thought this is how people feel before they die. I remember thinking I should go to the upper levels in case the building collapsed so that people could find my body. My prayer was that at least my body would be found in the rubble. When I got up there, the wind was really blowing strong, and it was impossible to stay. I went back to the third floor and waited it out.

Finally, it subsided, and once the winds got weaker, I got my bag. I left all the food and water I had with the people there. Just as I got out of the door, there were some dead bodies. I was eventually assigned to a crew picking up dead bodies around the city, together with six other people who volunteered. We had a dump truck that was lent to us by the local government. I think I recovered 78 dead bodies during that time. Three days later, I discovered that my friend Agit and

A CLIMATE MARCH IN TACLOBAN
BY DINO DIMAR, COURTESY OF A.G.SAÑO

his whole family, his wife and kid, and even his mom and dad were all killed. Their village was overrun with the storm surge, and they had to go out the door because the whole house was already flooded. When they got out, they were swept away by the waves. That's the last I heard of him. When I found out that Agit was dead, I asked the team leader if I could be transferred to a food relief mission because I knew I couldn't bear seeing Agit's dead body.

PUTTING A HUMAN FACE ON CLIMATE CHANGE

My older brother was part of the Climate Change Commission of the Philippines, which is an office of the Philippine president, whose mission is to negotiate during the UN climate summits. My brother was the head of the delegation at the time of one of the summits. Incidentally, it was happening around the same time as Super Typhoon Haiyan hit the Philippines.

He was given the chance to speak in the plenary at the UN summit, and he was given the microphone to address the UN and tell them about what happened to our father's hometown. Many climate experts in attendance said that his speech was probably the first time that they ever felt that there's a human face to climate change, because all they talk about is economics, fossil fuels, temperatures, emissions, but they never talk about the human cost of people actually dying or losing their communities.

LEARN MORE

Dakila: Philippine Collective for Modern Heroism
dakila.org.ph

A.G.'s work and message are brilliant, bold, and breathtaking. His larger-than-life murals have inspired people throughout the Philippines and around the world. His heart and empathy for others might be as big as the works he creates. We met a few years ago at an event where artists from all over the globe shared their work and how they use their talents to engage people around climate change and social and environmental justice. Everyone should aspire to be more like A.G., a world citizen fully aware of his responsibility to care for people and the planet. —DUDLEY EDMONDSON

A SIGN THAT READS "HOPE DIES LAST" NEXT TO A YOUNG BOY IN THE AFTERMATH OF TYPHOON HAIYAN
BY A.G. SAÑO

TACLOBAN CITY, THE PHILIPPINES, IN THE AFTERMATH OF TYPHOON HAIYAN
BY PHOTO BY A.G. SAÑO

WHAT?!

"While climate change is a global phenomenon, its negative impacts are more severely felt by poor people and poor countries."

—ORGANIZATION FOR ECONOMIC CO-OPERATION AND DEVELOPMENT

RESOURCES

IBRAHIM ABDUL-MATIN AND FATIMA ASHRAF

Green Deen: What Islam Teaches About Protecting the Planet, Kube Publishing, 2012

MAJORA CARTER

majoracartergroup.com/about

Reclaiming Your Community: You Don't Have to Move out of Your Neighborhood to Live in a Better One, Berrett-Koehler, 2022

CORINA NEWSOME

#BlackBirdersWeek

#BlackinNature

#RepresentationMatters

ROXXANNE O'BRIEN

National Community Reinvestment Coalition, ncrc.org

DEJA PERKINS

Fishin' Buddies, fishin-buddies.net

Shedd Aquarium, sheddaquarium.org

Lincoln Park Zoo, lpzoo.org

Brookfield Zoo, czs.org/Brookfield-ZOO/Home.aspx

QUETA GONZÁLEZ

Center for Diversity & the Environment, cdeinspires.org

DR. DREW LANHAM

Joy Is the Justice We Give Ourselves, Hub City Press, 2024.

Sparrow Envy: Field Guide to Birds and Lesser Beasts, Hub City Press, 2021.

The Home Place: Memoirs of a Colored Man's Love Affair with Nature, Milkweed Editions, 2017.

CHAD BROWN

Soul River Inc, soulriverinc.org

Love Is King, loveisking.org

AN EXPEDITION IN THE ARCTIC
BY DUDLEY EDMONDSON

RUE MAPP

Outdoor Afro, outdoorafro.org

Outdoor Afro Inc, outdoorafro.inc

Nature Swagger, Chronicle Books, 2023, chroniclebooks.com/products
/nature-swagger

CHRISTOPHER KILGOUR

Color in the Outdoors, colorintheoutdoors.com

NICOLE JACKSON

N Her Nature, instagram.com/nhernature

ALEX TROUTMAN

Critters of Georgia, AdventureKEEN, 2023. Alex is also the author of many
other books in the *Critters* series.

BlackAF in Stem, blackafinstem.com

SIQIÑIQ MAUPIN

Sovereign Iñupiat for a Living Arctic, silainuat.org

NIKOLA ALEXANDRE

Shelterwood Collective, shelterwoodcollective.org

TAMARA LAYDEN

The Center for Diversity & the Environment, cdeinspires.org
Xerces Society for Invertebrate Conservation, xerces.org

RICKY DEFOE

Water Legacy, waterlegacy.org
Antiracism Study Dialogue Circles, asdicircle.org

JASON HALL

In Color Birding, incolorbirding.org

ASHANEE KOTTAGE

LubDub Theatre Company, lubdubtheatre.com
ashaneekottage.com
kavayapress.com
instagram.com/kavayacollective

DR. LORENA RIOS MENDOZA

NOAA Sea Grant, seagrant.noaa.gov

DR. SEBASTIAN ECHEVERRI

BBC Earth Podcast, bbcearth.com/podcast
Crash Course Zoology, on YouTube
Spiders of the United States & Canada, Adventure Publications, 2024

CHARLIE "MACK" POWELL

P.A.N.I.C., facebook.com/PANIC.Bham

REVEREND EDWARD PINKNEY

Benton Harbor Community Water Council, bhcwc2.org
Black Autonomy Network Community Organization, bhbanco.org

SHARON LAVIGNE

RISE St. James, risestjames.org

A.G. SAÑO

Dakila: Philippine Collective for Modern Heroism, dakila.org.ph

ACKNOWLEDGMENTS

My wife, Nancy, has always been the primary person responsible for any level of my success I might have in the work I do. She's always believed in me and supported the things I've done to help promote diversity and inclusion in the outdoors over the last three decades.

Thank you to Stan Tekiela, who gave me my first real photography job decades ago. Stan believed in me, too, and hired me to photograph his many books on the flora and fauna of the United States. Stan gave me the chance I needed to fulfill my dream of being a professional nature photographer.

I also would like to mention my friend, the late Dr. Nina S. Roberts, PhD Professor, San Francisco State University, and Department of Recreation, Parks and Tourism. We met decades ago, during the writing of my first book *Black & Brown Faces in America's Wild Places*. In so many ways, she is responsible for my knowledge of diversity, equity, and inclusion issues in the outdoors. She introduced me to many of the people in that book and shared her knowledge with so many. Her absence is deeply felt by much of the BIPOC outdoors community today.

BY CHAD BROWN

ABOUT THE AUTHOR

Over the past 32 years, Dudley Edmondson has become an established photographer, author, filmmaker, and presenter. His photography has been featured in galleries and publications around the world, most recently in Australia and Italy.

In 2006, Dudley set out to create a group of outdoors role models for the nation's African American community by writing the landmark book *Black & Brown Faces in America's Wild Places*. During the Obama Administration, Dudley's book landed him an invitation to the White House for the signing of America's Great Outdoors Initiative.

Dudley currently lives in Duluth, Minnesota, and serves on the board for the Bell Museum of Natural History, as well as the Lessard-Sams Outdoor Heritage Council. You can learn more about Dudley via his website, dudleyedmondson.com.